Biocommunication and Natural Genome Editing

Günther Witzany

Biocommunication and Natural Genome Editing

Springer

Mag. Dr. Günther Witzany
Telos - Philosophische Praxis
Vogelsangstr. 18c
5111 Bürmoos
Austria

ISBN 978-90-481-3318-5 e-ISBN 978-90-481-3319-2
DOI 10.1007/978-90-481-3319-2
Springer Dordrecht Heidelberg London New York

Library of Congress Control Number: 2009938037

Cover illustration: Front cover image copyright Dennis Kunkel Microscopy, Inc.: Influenza A virus (RNA virus)

Printed on acid-free paper

Springer is part of Springer Science+Business Media (www.springer.com)

Preface

I wrote this book for biologists and those who are interested in both biological affairs in general and perspectives which integrate a large number of specialised biological disciplines.

The theory of biocommunication presented herein investigates signal transduction processes among cells, tissues, organs and organisms in bacteria, animals (corals and bees), fungi and plants in the light of the current available empirical data. Because life is the central focus of the life sciences, this theory will also focus on typical features of life as opposed to inorganic matter.

Because this field of investigation is based on the methodological primacy of a pragmatic action theory, the book may also be of interest to researchers of linguistics, communication sciences and sociology (e.g. plant sociology, animal sociology) who would welcome an overview of these highly specialised biological disciplines.

Current molecular biology as well as cell biology investigates its scientific object by using key terms such as *genetic code, code without commas, misreading of the genetic code, coding, open reading frame, genetic storage medium DNA, genetic information, genetic alphabet, genetic expression, messenger RNA, cell-to-cell communication, immune response, transcription, translation, nucleic acid language, amino acid language, recognition sequences, recognition sites, protein coding sequences, repeat sequences, signalling, signal transduction, signalling codes, signalling pathways, etc.*

All these terms combine a linguistic and communication theoretical vocabulary with a biological one. In this book I try to introduce an appropriate model to exemplify this vocabulary (which is used in biology all the time without people thinking about it), on the basis of explanation and understanding of a linguistic action, the great variety of communicative actions.

Many biologists are not very familiar with current definitions of 'language' and 'communication' in contrast to linguistics, communication science, pragmatic action theory, sociological theories. If we speak about (i) the three categories of signs, index, icon and symbol, (ii) the three complementary non-reducible levels of semiotic rules' syntax, pragmatics and semantics and (iii) communication as rule-governed sign-mediated interactions, it can easily be seen that all these categories are nearly unknown in biology, especially in molecular biology, cell biology, genetics and related disciplines.

Most biologists who use linguistic and communicative vocabulary to describe biological or genetic features do this according to their methodological self-understanding as empirical natural scientists. 'Language' and 'communication' are investigated in natural sciences behaviouristically or in the realm of formalisable procedures, i.e algorithms with information-theoretical, statistical or systems-theoretical conclusions.

In contrast with this linguistics, communication theory, semiotics and especially pragmatics, as well as action theory, have tapped knowledge about language and communication which was unimaginable 40 years ago. It is completely different from fundamental suppositions of older, mechanistic, behaviouristic and even information-theoretical definitions. In the light of this current empirical and theoretical knowledge it has become increasingly clear that the multiple levels of sign-mediated interactions which we call 'communication' cannot be explained or even sufficiently described by older models such as the 'sender-receiver' narrative or even based on the term 'information'. Even physicalistic or mathematical definitions of language as quantifiable sets of signs fail to describe key features of these phenomena.

The current theory on biocommunication does not want or is unable to replace empirical biology. Conversely, the recent quality of empirical biology is the prerequisite for a theory of biocommunication. Because natural sciences are not familiar with appropriate definitions of 'language' and 'communication' as both terms are in current action theoretical investigations this theory of biocommunication should be a complementary tool for e.g. biology. The theory of biocommunication then could act as a complementary tool in the interpretation of available empirical data concerning biological affairs.

The question is how we unite 'language' and 'communication' with biology. Why is it so comfortable and useful to operate with linguistic and communicative terms in biology? Are there any advantages for biology in research and teaching through a pragmatic theory of biocommunication? What is its traditional background, i.e. what is its place in the history of science? In this book I try to give answers to these questions. May I invite you to trace with me the roots of biocommunication?

Bürmoos, Austria Günther Witzany
August 2009

Acknowledgements

The first concept of the theory of communicative nature I published in 1993, a philosophy of biology in 'Natur der Sprache – Sprache der Natur. Sprachpragmatische Philosophie der Biologie' (Würzburg 1993) and its english and updated translation, 'Life: The Communicative Structure', followed some years later (Norderstedt, 2000). This new philosophy of biology was now recently applied in this concept of biocommunication. I have been privileged to hold excellent discussions with these outstanding experts, who encouraged me to develop this new approach. I want to thank Annemarie Pieper, Hermann Krings, Wilhelm Vossenkuhl, Rupert Riedl, Thure von Uexküll, Kalevi Kull, Don Favareau, Peter Harries-Jones, Frantisek Baluska, Peter Barlow, Frank Ryan and Luis Villarreal. For technical support I want to thank Wilhelm Hasenauer, Pierre Madl, Michael Stachowitsch, Karl Mayr and for everyday intensive communication Hiltrud Oman. Pierre Madl I want to thank for co-authorship of chapter 4, Biocommunication of Corals. I like to dedicate this book to the memory of my outstanding philosophy teachers Zeno Bucher and Beda Thum. Additionally I am grateful to Catherine Cotton, Ria Kanters and Ineke Ravesloot for excellent support during the preparation of this publication.

Contents

1 Introduction: Metaphysical and Postmetaphysical Relationships of Humans with Nature and Life 1
1.1 Metaphysical vs. Mythological Construction of Nature 1
1.1.1 Monistic-Organismic World Views 3
1.1.2 Pluralistic-Mechanistic World Views 4
1.1.3 Organic-Morphological World View 6
1.2 Delimitations Against Metaphysics 7
1.2.1 Linguistic Turn 8
1.2.2 Manfred Eigen's Adaptation of the Linguistic Turn to Biology 9
1.2.3 Deficiencies of Manfred Eigen's Depiction Theory of Language 11
1.2.4 Gödel's 'Incompleteness Theorem' and Real-Life Languages 13
1.3 The Roots of the Idea of an 'Exact' Scientific Language 15
1.4 Postmetaphysical Thinking: Pragmatic Turn 16
1.4.1 The End of Linguistic Turn 16
1.4.2 The Fundamental Status of Communicative Intersubjectivity 18
1.4.3 Evolutionary History: History of Rule-Governed Sign-Mediated Interactions 20
1.4.4 Biology in the Realm of a Theory of Biocommunication 20
1.5 Recent Applications of 'Language' and 'Communication' in Biology 21
1.5.1 Biolinguistics and Bioinformatics 21
1.5.2 Biosemiotics and Biohermeneutics 22
1.5.3 Biocommunication 23
1.6 The Structural Format of the Following Chapters 24
References 25

2 Plant Communication 27
2.1 Introduction: Multilevel Communication Competence of Plants 27
2.2 Chemical Vocabulary of Plants 29

2.2.1 Context-Dependent Auxin as Neurotransmitter, Hormone, Morphogenic Sign 29
2.2.2 Hormones 29
2.2.3 RNAs 31
2.2.4 Multiply Re-usable Components 31
2.3 Interpretation of Mechanical Influences 31
2.4 Transorganismic (Transspecific) Communication 32
2.4.1 Coordination of Defence against Pests and Injury . . . 32
2.4.2 Communicative Coordination of Symbioses 34
2.4.3 Vital Symbiosis of Plant Roots with Bacteria, Fungi and Animals 34
2.4.4 Viral Symbiotic Interactions 35
2.5 Interorganismic Communication 36
2.6 Intraorganismic Communication 37
2.6.1 Most Intercellular Communication via Plasmodesmata 37
2.6.2 Intracellular Communication 39
2.7 Plant Communication: Plant Neurobiology and the Emergence of Mind? 42
2.8 Conclusion 44
References 46

3 Communicative Competences of Honey-Bees 53
3.1 Introduction 53
3.2 Honey-Bees in the Colder Hemispheres 54
3.2.1 The Communication Process Behind the Founding of a New Colony 55
3.2.2 The Sign-Mediated Interaction of Foraging 57
3.3 Further Features of Honey-Bee Communication 58
3.3.1 The Types of Dances and Their Meanings 58
3.3.2 Forms of Communication Beyond Dances 59
3.3.3 Humans can Understand the Bee-Language 59
3.3.4 Dialects of the Bee-Language 59
3.4 Language and Communication in Bees: Context Determines Meaning 60
3.4.1 Foundation of a New Colony 61
3.4.2 Food Gathering 63
3.4.3 Dialects in Different Cultural Life-Worlds 64
References 64

4 Biocommunication of Corals 67
4.1 Introduction 67
4.2 Semiochemical Vocabulary of Corals 69
4.3 Interpretation of External Influences 70
4.4 Transorganismic (Trans-Species) Communication 71
4.4.1 Coordination of Defence and Regeneration 72
4.4.2 Communicative Coordination of Symbioses 73

4.5 Interorganismic (Species-Specific and Species-Related) Communication 74
4.6 Intraorganismic Communication 77
4.6.1 Intercellular Communication 77
4.6.2 Intracellular Communication 77
4.7 Conclusion 80
References 82

5 Biocommunication of Fungal Organisms 89
5.1 Introduction 89
5.2 Semiochemical Vocabulary of Fungi 91
5.3 Interpretation of Abiotic Indices 92
5.4 Transorganismic Communication 93
5.5 Biocommunication Among Fungal Species 95
5.6 Biocommunication Within Fungal Organisms 97
5.6.1 Intercellular Communication 97
5.6.2 Intracellular Communication 98
5.6.3 Unique Relationship Between Fungi and Viruses . . . 100
5.7 Conclusion 102
References 103

6 Bacteria Communication 109
6.1 Introduction 109
6.1.1 Biocommunicative Competences of Bacteria 109
6.1.2 Biofilm Organisation: Interpretation and Coordination . 111
6.2 Semiochemical Vocabulary and Communicative Goals 112
6.3 Transorganismic Communication 113
6.4 Interorganismic Communication 115
6.5 Intraorganismic Communication 117
6.5.1 Intracellular Communication 118
6.5.2 Bacterial Evolution and the Agents of Natural Genome Editing 119
6.5.3 Lytic versus Persistent Viral Life-Strategies 119
6.5.4 Bacteria as Biotic Matrix for Natural Genome Editing . 121
6.6 Conclusion 122
References 124

7 Natural Genome Editing Competences of Viruses and Virus-Like Agents 129
7.1 Introduction 129
7.2 Non-Coding Regulatory Networks 130
7.3 Major Viral Life Strategies 131
7.4 Examples of Diverse Viral Life Strategies 134
7.4.1 Virus Escape 134
7.4.2 Wall Off 134
7.4.3 Addiction Module: Reciprocal Interaction 134

7.4.4 Multiplicity Reactivation 134
7.4.5 Sexual Isolation 135
7.5 Pre-Cellular Life: Early RNA- and DNA-Viruses 135
7.6 Origin of the Eukaryotic Nucleus 136
7.7 Origin of the Adaptive Immunity 139
7.7.1 The Acquisition of a Complex New Phenotype 140
7.7.2 Ancestral Origin of an Adaptive Immune System . . . 142
7.8 Evolution of Placental Mammals 142
7.9 Conclusion 143
References 145

8 How Bacteria Escaped Selection Pressure of the Early RNA-World 149
8.1 Introduction 149
8.2 From Pre-Cellular RNA-Copying to RNA-Coding 150
8.3 Communal Evolution: From LUCA to LUCAs 151
8.4 Old but Good: Current Competences from an Ancient World . . 152
References 154

9 Viral Origins of Telomeres and Telomerases 157
9.1 Introduction 157
9.2 Different Molecular Syntax of Telomere Sequences 158
9.3 Telomere Replication in Most Cases by Telomerase 159
9.3.1 Reverse Transcriptases and Mobile Elements 159
9.3.2 Roles of Reverse Transcriptases in Natural Genome Editing 160
9.4 Telomeres are Characteristics of Eukarya 162
9.5 Agents of Natural Genome Editing 163
9.6 Superficial and Deep Grammar in Eukaryotic Genome Content . 164
9.7 Conclusion 165
References 166

10 Real Life-World of Noncoding RNA-Species 171
10.1 Introduction 171
10.2 Genetic Text-Sequences Function Similar to any Natural Language 172
10.3 Cellular DNA Nucleotide Sequences as Viral Life Habitat . . . 173
10.3.1 The Persistence of the Eukaryotic Nucleus 174
10.4 Viral Agents as Genetic Editors 175
10.4.1 Persistent Viral Life Strategies Change Genetic Host-Identities 176
10.4.2 Former Competing Genetic Parasites Built Addiction Modules 176
10.5 Competent Regulators of Gene Expression 180
10.5.1 Identification and Regulation by microRNAs and siRNAs 180
10.5.2 Non-coding RNAs act as Ribonucleoproteins 183
10.5.3 Small Nuclear and Small Nucleolar RNAs 184

10.5.4 Currently Identified Roles of Small Nucleolar RNAs . 185
10.5.5 The tRNA Consortium 187
10.6 Reciprocal Interacting Agents 188
10.7 Conclusion 190
References 191

11 Outlook 197
11.1 From Mechanistic Biology to Biocommunication 197
11.2 Three Kinds of Signs in Biocommunication 199
11.3 Context Determines Meaning 200
11.4 Living Nature and Non-living Nature 200
11.5 Biocommunication Defines a Biotic 'Plus' 201
11.6 The Advantages of Biocommunicative Biology 202
11.7 Linguistic and Communicative Competences in Non-human Nature 202
11.8 Complementary Roles of Linguistic and Communicative Competences 203
11.9 New Qualities for Future Decisions 203

Index 205

Chapter 1
Introduction: Metaphysical and Postmetaphysical Relationships of Humans with Nature and Life

Abstract First, I offer a short overview on the classical occidental philosophy as propounded by the ancient Greeks and the natural philosophies of the last 2000 years until the dawn of the empiricist *logic of science* in the twentieth century, which wanted to delimitate classical metaphysics from empirical sciences. In contrast to metaphysical concepts which didn't reflect on the language with which they tried to explain the whole realm of entities empiricist *logic of science* initiated the end of metaphysical theories by reflecting on the preconditions for foundation and justification of *sentences* about objects of investigation, i.e. a coherent definition of language in general, which was not the aim of classical metaphysics. Unexpectedly empiricist *logic of science* in the linguistic turn failed in the physical and mathematical reductionism of *language* and its use in *communication*, as will be discussed below in further detail. Nevertheless, such reflection on language and communication also introduced this vocabulary into biology. Manfred Eigen and bioinformatics, later on biolinguistics, used 'language' applied linguistic turn thinking to biology coherent to the *logic of science* and its formalisable aims. This changed significantly with the birth of biosemiotics and biohermeneutics. At the end of this introduction it will be outlined why and how all these approaches reproduced the deficiencies of the logic of science and why the biocommunicative approach avoids their abstractive fallacies.

1.1 Metaphysical vs. Mythological Construction of Nature

Linguistic and communicative vocabulary as a crucial tool in scientific foundations and the methodology of philosophy of science has been in use for 70 years. Before this time, empirical descriptions of non-living nature and even living nature were derived from metaphysical constructions with a long and complex history embracing the most prominent thinkers in occidental philosophy. All of them tried to give answers to the classical antinomies which derived from the Athens school of Greek philosophy. Before I give a short reconstruction of the metaphysics of nature I want to give a synopsis of what the metaphysical thinking opposed. It was a central

G. Witzany, *Biocommunication and Natural Genome Editing*,
DOI 10.1007/978-90-481-3319-2_1, © Springer Science+Business Media B.V. 2010

paradigm shift in human history: the change from a myth-based self-understanding with its focus on cultus and ritus and the strict hierarchy of norms and traditions within which a tribal society was interwoven. Pre-metaphysical hierarchies are mythology-based forces of creation. Nature was speaking to humans as animals and plants, natural forces as thunder, wind, water and fire. The order of the world was self-evident. Animals were not ranked inferior to humans but equally. As animals are different, so are humans different. The self-evident order of the world is a cosmocentric law which rules over animals and humans. The myth of the change of nature before humans to nature with humans does not resolve the status of nature without humans. The mythology of pre-metaphysical tribal societies suppresses destruction of nature definitively. Nature is a kind of holy being within which non-holy humans are embedded. Therefore humans have to act accordingly.

Humans in pre-metaphysical tribal societies were not only ecological experts. Their educational systems were holistic ones, each member being required to be familiar with their surroundings such as plants, animals, climate, annual cycles and repetitions, interdependencies of the inner and outer nature. Each member of this kind of human society was also familiar with the ethics and norms of tribal traditions in social affairs.

A different relationship with nature was constructed in the metaphysical thinking of the classical Greek philosophers. At the basis of the western occidental-modern world view and technical-scientific modernity we can find Greek cosmology. Their competing metaphysical world views are extensively developed constructions according to logics and methodology which offer completely different answers to questions of the myth-based lifeworld.

The change from a natural being into a society-based being is an irreversible process. The division of survival of society, of the survival of non-human nature, indicates a newly-derived hierarchy in which culture (i.e. inner nature of humans) has primacy in opposition to nature (outer nature). The subordination of society to an omnipotent creator god and his plan of creation are followed by the subordination of non-human nature to the human one. The hierarchy is strict and structured: the primacy of gods, followed by humans and, last, the rest of nature. The age of unity between human mankind and nature is broken irreparably. The order of the world is no longer self-evident but offered by god and supernatural. God is thinking prime pictures whose depictions are manifested in a great variety here on earth and in the cosmos (Capelle 1968).

The invention of the *general term* as crucial tool of the technique of abstract thinking divides metaphysical interpretation of nature clearly from the pre-metaphysical one.

The rationalisation of world views occurred in parallel with differentiation and complexity of writing. Development and practice of the technique of writing liberated transport of tradition from the ancient practice of vocal traditions. Now it was possible to read about the history and myth of tribal societies even if they were far away or no longer present. Metaphysical philosophy of nature from now on had to answer the questions of classical antinomies, the relation between the *whole and its parts* and between (statical) *being* and (dynamical) *becoming*.

There are three mainstream conceptions within metaphysical philosophy of nature in the occidental tradition of philosophy of the last 2000 years. All philosophical conceptions of the last 2000 years are part of one of these mainstream paradigms.[1] We differentiate:

- Monistic-organismic world views
- Pluralistic-mechanistic world views
- Organic-morphological world view

1.1.1 Monistic-Organismic World Views

The main principle of all monistic-organismic world views is holism (all is one). What we experience as a broad variety of things and processes constituting this world is attributed to one main principle. The multitude of beings is within this world view deduced from one driving force. Cosmos is a whole, the many things are epiphenomena which seem to be many but in reality are only parts of the whole. Inside and outside are two aspects of the one and whole reality. What seem to be many are only different moments of the one and all. According to different wholes in this monistic organismic world view (Life, soul of the world, world-mind, god) there is a differentiation between a physical, metaphysical or pantheistic monism. If the main principle is life we speak about hylozoism, if it is the soul, panpsychism, if it is divine, pantheism. In all of these monisms there is one and only one main principle which is behind all things. In the history of philosophy we can differentiate different developments of these monisms such as pre-Socratic hylozoism (Thales of Miletus, Anaximenes, Heraclitus), cosmic pantheism of the Stoa (Seneca, Epictetus, Marcus Aurelius, Cicero), pantheistic emanantism of Plotinus (Ammonius Sacca, Plotinus, Scotus Eriugena and later on Spinoza, Hegel), aesthetic pantheism of Giordano Bruno and his monadology, which was further developed by Leibniz and the pre-critical Kant in his metaphysical dynamism, and later on Hamann, Kierkegaard, Schelling, Goethe, Novalis, Hölderlin, Rilke, Steiner.

In a certain sense, this monistic world view is exemplified also in rationalism with its objection that the whole world can be imagined as and integrated within one objective and logical system which we must solely investigate long enough to integrate all things into this one and only system, as thought by Spinoza. This thinking also attracted Hegel. The god of Hegel is living and organismic and emerges as world through dialectical processes of birth, death and next level of being. In its organismic variation we find monistic evolutionism in Clifford, Huxley, Darwin and Spencer. One law determines the whole universe. This absolute unifying law is the law of development. Differentiation and Integration are the everlasting potentials of this law. The Emergentism of Samuel Alexander postulates the one and only world matter which is the material out of which all things are formed. The many parts and

[1] The outline of the three metaphysical world views follows Zeno Bucher (1982). Natur, Materie Kosmos. Eos, St. Ottilien.

processes are events which are emerging out of this world matter which is at the last identical with god.

Another much younger philosophy is the panvitalistic metaphysics of France with Maine de Biran and Bergson. From its strict anti-mechanistic and anti-rationalistic view they propagated a self-enforcing power of all living, or as Bergson called it, the Bios, the principle of creation in the whole reality which is the driving force of the whole universe. This vitalism is integrated within certain other holisms which tried to unify this world view with modern natural science knowledge and such proponents as Haldane, Meyer-Abich, Wheeler, Whitehead, Bohm, Capra.

Another kind of organic monism is the dialectical materialism of Engels which is a counterpart to Hegel's idealistic monism. The whole and the one is more than the sum of its parts. The parts per se have no value, only in sum the whole is the main value.

The monistic-organismic world view is present also in twentieth-century physics. Searching for the last common invisible matter, or the elementary parts of all matter or the last and one formula (Stephen Hawking) through which all can be explained or which represents the ultimate law of all being, are variations of 'all is one' – metaphysics. The particles on the subatomar level are not parts by their own. They are parts which are all constituted by lower parts and smaller parts and at the last they are quantums, quantum parts (Heisenberg 1973) or, as Einstein noted 1950, electrical field densities which we see as corpuscles but in reality are condensed out of a universal field of energy.

The driving force of these monistic-organismic world views derives from both the presumption of the unification of thinking and being (without language's critical reflection) and a kind of idealistic rationalisation of experiences such as separation, transitoriness, contradiction, fear of reality and the new, unexplainable. The many parts we experience are at the last all in one, a common principle, the last and ultimate law and formula or in its theistic variation basics of all religious social orders (Wittgenstein 1975: 80e).

1.1.2 Pluralistic-Mechanistic World Views

In strict opposition to the monistic-organismic world views there are the pluralistic mechanistic ones. Their main principle is: all is endless plurality (all is many). In contrast with the holistic one and only of monisms, in the pluralistic all is built of indefinite numbers of corpuscles, smallest parts. If we look at or experience things, persons, objects, they seem to be entities, but in reality they are the sum of these smallest corpuscles. These last smallest entities are unchangeable and everlasting. A real becoming out of nothing, i.e. a real de novo emergence which means a movement from not being into being may be a construction in our consciousness but has nothing to do with any reality. These ultimate single corpuscles can be brought into forms or can even be mixed but this does not change anything within them. In their outer nature they can be moved and change their relation to one another but in their substance they are unchangeable. Any movement is caused from outside and is purely mechanistic. Parmenides was one of the first thinkers to propound this world view. For him movement is also an illusion because it is a line-up of the smallest unmoved statical state of things.

A hundred years later the atomistic school changed the philosophy of Parmenides in one crucial aspect. Leukippus and Democrites now believed the experience of multiplicity, changeability and movement to be reality. The hylomorphistic conception of Aristotle damaged this world view until it was revived by Pierre Gassendi in 1649. He constituted the philosophy of mechanistic atomism in a new way. Robert Boyle described this mechanistic atomism. He observed that in contrast with older forms of atomism matter is an assembly of different basic elements. His philosophy focused on investigations and research into these basic elements. A hundred and fifty years later Proust and Dalton postulated real atoms in the so-called law of constant and multiple proportions.

Then the term *molecules* was developed and in the shift to the twentieth century it became increasingly clear that atoms are not atoms, because they are constituted by a variety of dynamic entities which can emerge as corpuscles or waves. This contradicted the term atom fundamentally. Although atomism was shown to be a misinterpretation of nature, Mechanism as mechanism has been a successful model until today.

René Descartes observed earlier that the concept of indivisible corpuscles is dubious in principle. Rationalistic investigation can experience only mathematical relations and the reality of matter can only be viewed as proportion and dimension. We can only understand machines because all their functions can be reconstructed by investigation of the function of their parts. These parts of the world machine first are thought by god and later on produced. The parts of being within Descartes's thinking are purely dimensional and equal. Behind these qualities of the ultimate parts there is nothing else. They can be differentiated only in their size, geometry and configuration. Every phenomenon in the cosmic universe is a configuration and local movement of these parts. Additionally all living beings function in the same way, purely mechanistically.

Descartes's strict mechanisation of everything within nature became a broad mainstream world view. The principles of mechanics were adapted to whole physics and as result this kind of physics became the basic science of all empirical research and investigation. A late player was Newton, whose philosophy was founded on mechanistic principles. According to Newton, ultimate particles are created by god as massive particles. From this, the next step was the apodictic mechanism of LaPlace, who stated that every single status within the world and cosmos is a strict effect of the foregoing causes. If there were a mind which could oversee all forces which are existent in nature it would be able to predict every future development out of this knowledge, because everything happens according to strict mechanical laws (the LaPlace demon).

Then came a state of universal determinism: the state of every closed system at a certain moment determines the following development for all time. The whole world as well as the universe is a big machine, which is constituted by an infinite quantity of parts, all of them underlying strict natural laws. With strict rational thinking nature has to be analysed in minutest detail until all parts of nature are part of scientific knowledge. Then sometimes mechanical nature can be reconstructed completely and even optimised, unlike real nature.

This is also valid for all living beings including humans, especially the human mind. This was a basic conviction also of Dubois-Reymond, one of the main mentors of Sigmund Freud. In the twentieth century, Rudolf Carnap was also convinced that the psychological features of humans are a bundle of physiological mechanistic single processes and should be explained mechanistically.

1.1.3 Organic-Morphological World View

In between these two completely contradictory world views there is a third world view which was worked out by Aristotle and later on by Thomas Aquinas, and has been further developed in the twentieth century by the school of Neo-Thomists like Nikolai Hartman, Aloys Wenzl, Hedwig Konrad-Martius and others. The starting-point of the so-called organic-morphological world view is the theory of levels, which includes a categorisation of the delimitations and differences between these levels. Hartman distinguishes four levels of being: the material, the vital, the psychic and the mental. These levels differ in their stages on the way to perfection which depend on the translation of potentiality into actuality. In the level of the material, the potentiality is dominating, actuality is less, i.e. the real matter of the world behaves according to natural laws, e.g., in the case of nuclear technologies, much more actuality can be processed out of single atoms. In a nuclear chain reaction actuality is nearly indefinite whereas potentiality approaches zero. The higher level integrates the lower one, although the lower one is not dissolved but gets a new function within the higher one. This means reality is constituted by many and ultimate smallest particles which develop in real processual reality into different forms which unite to become such things as bodies of living beings. This organic morphological world view strictly contradicts the monistic-organismic as well as the pluralistic mechanistic world view. The relationships between these levels are determined by a set of laws:

1. Law of autonomy: according to Hartmann, each layer of being is autonomously structured and the genesis of this autonomy cannot be fully derived from the next lower level. The mental level is therefore independent of the psychic level, the psychic of the vital, and the vital of the inorganic. This does not necessarily mean that the mental level lacks the psychic level, the psychic level lacks the vital one and the vital lacks inorganic elements. Rather it emphasises that each of these levels is characterised by features that can be found here and only here. Within this law of autonomy there are two subordinate laws. (a) The law of novelty: in each higher level, features appear which are lacking in the next lower level. These features represent a novelty, something new compared with the lower level. Such new features are neither a logical consequence in the development from the lower to the higher level nor can they be fully derived from the former. (b) The law of modified, recurrent features: the laws of the lower level reappear in the higher level, never vice

versa, but in a modified manner. Specifically, the laws of the lower are structurally and functionally integrated into the higher. For example, the laws of the inorganic level recur in the vital level, but under organisational principles of the vital level, i.e. in a constellation unknown in the inorganic level.

2. Law of dominance: the laws specific to one level do not merely govern that layer. Within the overall organism, every higher level acts on all levels below it, without dismantling or negating them. Humans, for example, possess a vegetative nervous system whose function is largely independent of mental activity. This mental activity, however, can influence the psychic state and, by destabilising it, e.g. in extreme stress situations, have an effect on the vegetative nervous system.
3. Law of dependence: each higher level is neither poised above nor determined by the lower ones, although a certain dependence does exist. The mental level functions on the basis of the psychic, this on the vital, and the vital in turn on inorganic substances. In the case of comatose patients, the vital level and the vital organisation of the inorganic matter comprising the body continue to function, but the psychic and mental levels are silenced.
4. Law of distance: owing to the new, defining quality of a level of being, Nikolai Hartman recognises a 'metaphysical discontinuity' rather than actual transitions between these levels. While representatives of approaches based on continuity theories have always postulated such transitions, no actual transitions have been found or convincingly reconstructed in the field of palaeontology. According to this law, nature, and even evolution, progresses in discrete steps.

In contrast with the former two most prominent world views with their 'all is one' or 'all are part', in this different worldview being is a kind of processual reality with developmental stages from simpler to more complex structures. Also in contrast with the former conception, the occurrence of novelty which did not exist before is a special feature of being and cannot be logically or ontologically deduced from former stages. Movement, development, changeability from the littlest inorganic parts up to the human mind is an inherent potentiality of being and not a mere epiphenomenon or mixture of unchangeable smallest beings. The differences between the organic morphological world view and the monistic-organismic and pluralistic-mechanistic world views are fundamental and unbridgeable.

1.2 Delimitations Against Metaphysics

All schools of philosophy from antiquity and the classical Greek age up until the twentieth century tried to solve the classical problems of antinomy, i.e. (i) the relationship of the whole and its parts and (ii) the (statical) being and the (dynamical) becoming. The short overview of the philosophical conceptions, their tendencies and motifs described a kind of philosophy which was to be strictly avoided by the philosophy of science called logical empiricism (neo-positivism) and later on critical rationalism. The whole dictionary and language game played in these metaphysical languages was a real nightmare for the proponents of the project of 'exact science'.

They were convinced they could find a language which could both exclude metaphysical language, inexact terms and apodictically-claimed truth and for the future express empirical sensory data unambiguously and definitively.

For logical empiricism metaphysical questions do not have any subject and therefore replace this kind of philosophy by the primacy of empirical scientific knowledge, materialism and naturalism. As we will see later, both the idealistic tradition and it materialistic counterpart and even empirism share classical metaphysical positions such as (i) their claim of being an original philosophy and (ii) the identification of being and thinking. The latter one particularly constructs an inner relationship between thinking and being: as we are thinking, being also functions. The general, the necessary and the supratemporal can be found also in their terms. Empirism and Nominalism identified this self-misunderstanding. They resolved this misunderstanding in a multitude of entities without qualities. Only by the sensory organs of feeling subjects can these entities be mentioned and then be reconstructed within their imaginative apparatus.

Modern empirism wanted to be freed from these metaphysical implications by a substantial and fundamental critique of metaphysics. Therefore the only serious value for science is the rationality of the methods of scientific knowledge, i.e. the formalisable expression of empirical sentences. This is strict objectivism, which restricts itself to a pure observer perspective that confirms its observations by techniques of measurements and subsumes reality in the formalisable depiction of these measurements. Between metaphysics and objectivism there is an unbridgeable gap: what can be found empirically and described as formalisable exists. Outside these criteria everything can be believed but is not the subject of exact sciences. From now on objectivity is the main agenda of natural sciences, subjectivity the subject of human sciences. The interesting and ambitious programme of logical empiricism started as no scientific discipline started before: by a fundamental critique of the sentences with which we describe observations and those with which we construct theories.

This scientific approach was a fundamental shift in the history of philosophy. In the main focus were not the things, the world, the being but conversely the medium in which we describe our opinions, impressions, experiences, the language itself.

1.2.1 Linguistic Turn

To delimitate exact scientific sentences from inexact sentences as they occur in philosophy and theology, the school of logical empirism (Carnap 1931a, 1931b, 1934, 1939, 1956, 1966; Neurath 1932; Gödel 1931, and later on Russell 1940; Tarski 1966) at the beginning of the 1930s tried to construct a formalised language of exact sciences according to the young Ludwig Wittgenstein and the theoretical construction which he outlined in *Tractatus Logico-Philosophicus* (Wittgenstein 1959). With this formalised language of exact sciences it should be possible to outline empirical results of experimental research exactly and without ambiguity. This means that every sentence with which observations are described as well as sentences which are used to construct theories must fulfil the criterion of formalisability, i.e. they must be expressible as mathematics. Only sentences which fulfill this criterion should be claimed as (termed as) scientific. Sentences which would not be formalisable have to be excluded from science because they are not scientific sentences. By this procedure natural science should be installed as exact science. Because the world functions exclusively according to the laws and principles of physics, this world can be depicted only by sentences of mathematics which are able to express physical reality in a one-to-one manner. Natural laws expressed within the language of mathematics, i.e. formalisable, represent the inner logic of nature. The central part and most important element of language therefore is the syntax, because only by the logical syntactic structure of language is it possible to depict the logical structure of nature. Language as depiction of the natural laws of reality therefore must be formalisable in all its aspects. Because language therefore is seen as a quantifiable set of signs it can be expressed also in binary codes (1/0). Meaning functions therefore are deducible solely from this formal syntactic structure.

Similarly to this model of language, cybernetic system theory and information theory investigate the empirical significance of scientific sentences out of a quantifiable set of signs and, additionally, out of the information transfer of formalised references between a sender and a receiver (sender-receiver narrative). Information-processing systems therefore are quantifiable themselves. Understanding information is possible because of the logical structure of the universal syntax, i.e. by a process which reverses the construction of meaning. Because of this theorem, information theory is also a mathematical theory of language (Shannon and Weaver 1949; Turing 1950). Both constructions are founded on the assumption that reality can be depicted in a one-to-one manner only by formalisable procedures, i.e. formalised sentences. Exact sciences means correspondence of thinking and being. Manfred Eigen adapted these models for biology in the last third of the twentieth century in the description of the genetic code as a language-like structure (Eigen and Winkler 1975).

1.2.2 Manfred Eigen's Adaptation of the Linguistic Turn to Biology

Manfred Eigen compares human language with molecular genetic language explicitly.[2] Both serve as communication mechanisms.[3] The molecular constitution of genes is possible, according to Eigen, because nucleic acids are arranged according

[2] 'Speech, communication, reading and comprehension on this level mean binding (=recognising) the complementary molecular building blocks (=language symbols) and linking them into a macromolecular ribbon (=text)' (307).

[3] 'Each language primarily reflects the characteristic features of the respective, underlying communication machinery' (313).

to the syntax and semantics of this molecular language. [4] Even the amino acid sequences constitute a linguistic system.[5] Through this comparison Manfred Eigen follows the depiction theory of language within the tradition of empiricism, logics, mathematical language theory, cybernetic systems theory and information theory.

The world behaves according to physically determinable natural laws. These natural laws can be expressed only by using the language of mathematics. The formalisable artificial language of mathematics is alone capable of realistically depicting these natural laws. Language in its fundamental sense is language as a formalised sign language. The natural laws are explications of the implicit order of mathematics and nature. Mathematical language depicts this logical order through the logical structure of the linguistic sign system. The essential level of rules of a language therefore is the syntax. Only through the syntax does the logical structure of a language as a depiction of the logical structure of nature come to light. Because both the identity of the logical order of the language in its syntax and the logical order of nature can be expressed in mathematics, this language is quantifiable and can be expressed in binary codes (1/0).

The semantic aspect of language initially comprises an incidentally developed or combined sign sequence, a mixture of characters, which only gain significance in the course of specific selection processes. The linguistic signs are variables whose syntax is subject to the natural laws governing the sign-using brain organ. The brain of humans, for example, is endowed with these variables and combines them to reflect synapse network logics. The variable sign syntax of the brain then must be filled up with experiences of a personal nature and thus constitutes an individualised evaluation scheme.

In messages between communication partners, one side encodes the message in phonetic characters. The receiver must then decode and interpret the message based on empathy and personal experience. Understanding messages shared between sender and receiver is largely possible because the uniform logical form – a universal syntax – lies hidden behind every language.[6]

The function of that organ which syntactically combines the language signs according to its own structure most closely corresponds in Eigen's opinion to cybernetics, i.e. the theory of information-processing systems (while abstracting the manner of its realisation). Functional units like the central nervous system, brain or even macromolecules consist of a definable, limited number of elements and a

---

[4] 'The relative arrangement of the individual genes, the gene map, as well as the syntax and semantics of this molecular language are (...) largely known today' (207).

[5] 'Although the active center – the actual three-dimensional word correlate of the protein language – comprises no more characters than the number of verbs in spoken language, the protein molecule must unite a total of between one to five hundred chain elements within itself in order to form such an active center, each one of these molecules represents a particular task and one could describe the enzymes as the ‚verbs' of the molecular language'.(305) '(...) All the words of the molecular language are combined to a meaningful text, which can be broken down into sentences' (305).

[6] '...sentence structures, if we disregard the specific peculiarities of the individual languages, exhibit parallels that indicate a universal regularity evidently originating in the organization of the human brain' (301).

limited number of relationships between these elements. These systems, along with their description by means of a language, are depictions of a reality, structured by natural laws. Since both the logic of the describing and that of the theory constructing language correspond with the logic of the system, the relationship between these elements of the system can be represented in an abstract, formal and unambiguous manner.

From the perspective of man as a machine, humans clearly represent an optimal model: they fulfil all those preconditions for constructing algorithms that a conventional machine cannot deliver, i.e. criteria for information evaluation based on the real social lifeworld.[7] Humans, and all other biological systems, resemble a learning machine capable of internally producing a syntactically correct depiction of the environment by interacting with this environment, of correcting this depiction through repeated interactions and thus of changing their behaviour according to the environmental circumstances. Such learning systems are able to continuously optimise their adaptability.[8]

The differences between nucleic acid language and human language stem from the continuous developmental processes of biological structures, based on the model of a self-reproducing and self-regulating automaton that functions as realisations of algorithms.[9] This enables the steady optimisation of problem-solving strategies in organisms, eventually leading to the constitution of a central nervous system, a precursor ultimately giving rise to the brain and its enormous storage and information-processing capacity. Language enables the implementation of this evolutionary plan (from the amoeba to Einstein): this medium forms, transforms, stores, expands and combines information.[10]

1.2.3 Deficiencies of Manfred Eigen's Depiction Theory of Language

Even formal systems are not closed, as Eigen suggests, nor are they principally fully determinable. Furthermore, language is the result of communicative interactions

[7] 'Nature, through the development of receptors that register environmental signals and through the development of nervous systems that can process and store such signals, has found a more economic way' (225).

[8] 'A specific operational task of the von Neumann automaton is self-reproduction. The first model from the year 1950 was entirely realistic in its conception: the machine runs back and forth in a huge spare-parts warehouse and compiles the components necessary for its own replication. Most importantly, it also reproduces its own construction plan or blueprint. Its progeny should, after all, also be equipped with the self-reproduction capability. Herein lies the possibility to perfect the von Neumann automaton, an idea that has long been taken up by theoreticists: selective alteration of the program enables continuous improvement and an expanded range of application in the sense of Darwinian evolution' (216).

[9] 'In principle, the automaton is capable of carrying out any desired calculation' (217).

[10] 'At any rate one can say that the prerequisite for both great evolutionary processes of nature – the origin of all forms of life and the evolution of the kind – was the existence of a language' (314).

in dialogue situations rather than the result of constitutive achievements of the individual persons. Communicating with one another, sending messages, understanding expressions is not a private coding and decoding process, but rather an interpretation process arising from a mutual adherence to rules by communicating partners who agree on the rules.

The ability to abide by these rules is innate, the skill in complying with particular rules is acquired through interactions and relies on norms of interaction to utilise words in sentences, i.e. a linguistic competence. Information cannot generally be quantified as message content: statements made by social individuals in situational contexts are not closed and, thus, are generally not fully formalisable. The attempt to construct a purely representational language is doomed to failure because formal artificial languages do not exclusively contain terms that are unambiguous. This pertains to terms that cannot be confirmed through observation. Specifically, scientific statements are not attributable to immediate sensory experience, i.e. *the language game used to describe observations does not mirror the brain activity during the perception of reality.*

A world-depicting exact language must remain a mere postulate because it cannot logically substantiate itself. Too many theoretical concepts, too many scientific criteria that are generally not formalisable (e.g. 'progress in the cognition process', 'practicability', etc.), point to the limits of formalisability. The very identity between artificial language and its form renders it incapable of reporting on itself, something that presents no problem when informal speech, i.e. everyday language, is used. Language is an intersubjective phenomenon which several individuals can share, alter, reproduce as well as renew the rules of language usage. The basis and aims of this usage are defined by the real social lifeworld of interacting life-forms. The user of a linguistic sign cannot be comprehended according to the speaker-outside world model. Rather, this requires reflection on the interactive circumstances to which the user has always been bound, circumstances which provide an underlying awareness enabling him/her to understand statements made by members of the real lifeworld. The user of formal artificial languages – before appreciating the purpose of the usage – has also developed this prior awareness in the course of interactive processes with members of the real social lifeworld.

Speech is a form of action, and I can understand this activity if I understand the rules governing the activity. This means I can also understand an act that runs counter to the rules. Everyday language usage reflects everyday social interactions of the constituent individuals. The prerequisite for fully understanding statements is the integration of the understander in customs of social interaction and not merely knowledge of formal syntactic-semantic rules. A prior condition for all formalisations in scientific artificial languages is a factual, historically evolved, communicative experience. This very precondition becomes an object of empirically testable hypothesis formation in Eigen's language model. At this point, however, Eigen's model becomes paradoxical because he seeks to grasp theoretically language with tools that are themselves linguistically predetermined.

Eigen's language model, which is rooted in information theory, clearly reveals that Eigen equates the form of theory language with the form of language used to describe reality (experience). This implies the equation of formalised scientific languages with the language used to describe observations. Previous attempts to specify all the rules governing the translation of every term in theory-language into terms of observational languages have been unsuccessful. Not all concepts of theory language can be transposed into concepts of the observational language.

1.2.4 Gödel's 'Incompleteness Theorem' and Real-Life Languages

A similar situation (but more than 40 years before Eigen) is encountered in the attempt to absolutise mathematics as that pure formal language whose every ramification might become fully transparent. This led Gödel to formulate the *Unvollständigkeitssatz* (incompleteness theorem) in his work *Über formal unentscheidbare Sätze der principia mathematica und verwandter Systeme* (1931). Gödel investigated a formal system by applying arithmetic and related deduction methodologies. His aim was to convert metatheoretical statements into arithmetical statements by means of a specific allocation procedure. More precisely, he strove to convert the statements formulated in a meta-language into the object language S by using the object language S. This led Gödel to two conclusions:

1. On the assumption that system S is consistent, then it will contain one formally indeterminable theorem, i.e. one theorem is inevitably present that can be neither proved nor disproved within the system.
2. On the assumption that system S is consistent, then this consistency of S cannot be proved within S.

The question of determinability and calculability is closely allied with the algorithm concept, whereby Eigen seems to postulate that algorithms are not only concepts of theoretical language, but also depict (decision-) behaviour in the realm of biology and, therefore, are amenable to empirical analysis. Indeed, he is convinced that everything can be represented in the form of algorithms and can thus, in principle (after sufficiently thorough analysis), be determined. Yet Eigen never puts this to the test, i.e. he never states the conditions in which a branch of mathematics would be indeterminable. Namely, a field of formalised artificial language is indeterminable when no algorithm can be provided to help one to decide – for a particular formula of a formalised artificial language and involving a finite number of steps – whether this formula is universally valid or not.

Today, several branches of mathematics are considered indeterminable. Herein lies the consequence of this indeterminability theorem for the automaton theory of A. Turing and J. v. Neumann: a machine can principally calculate only those functions for which an algorithm can be provided. Functions lacking an algorithm are not calculable.

Every cybernetic, self-controlling machine is the realisation of a formal system. Eigen assumes that the evolution of self-reproducing and self-organising organisms represents the realisation of the syntax of a universal language underlying the order of the world. This universal syntax, as a representation of mathematically expressible reality, is also the formal basis for the evolution of these organisms. For each of these machines, as in the case of every organism, there must be an indeterminable formula.

It is precisely by means of a non-formal language that this formula can be shown to be true or false; this non-formal language is the very tool that enables the language itself to be discussed. The machine is unable to do this because no algorithm is available with which a cybernetic machine can determine its underlying formal system. Systems theory is principally unable to fulfil the demands that Eigen places on it.

The fact that the paradoxes arising within a formal language cannot be solved with that language led to a differentiation between object language and meta-language. Nonetheless, paradoxes can also appear within meta-language; these can only be solved by being split into meta-language, meta-meta language and so forth in an infinite number of steps. This unavoidable gradation of meta-languages necessitates resorting to informal speech, developed in the context of social experience, as the ultimate meta-language. It provides the last instance for deciding on the paradoxes emerging from object- and meta-languages. Neither the syntax nor the semantics of a system can be constituted within that particular system without resort to the ultimate meta-language.

The ambition to provide logic and mathematics with a priori validity is no longer tenable: an unambiguous linguistic foundation of science, one beyond further inquiry and supporting itself through direct evidence, cannot be secured. Language proves to be a perpetually open system with regard to its logical structures and cannot guarantee definiteness from within itself. This is the very conclusion that Eigen disputes with his language model. To summarise this chapter:

- There can be no formal system which is entirely reflectable in all its aspects while at the same time being its own metasystem.
- Concrete acts and interactions are basically unlimited in their possibilities. There will always be lines of argumentation that lie outside and have no connection with an existing system. Basically, every system can be transcended argumentatively. Newly-emerging language games and rules may develop as novel structures which are foreign to previous systems and not merely a further step in a series of prevailing elements. These very discontinuities enable totally new language applications.
- The ultimate meta-language, informal language, provides indispensable evidence about the communication practice of subjects in the real environment; the operator of formalisations is itself an integral part of this. Reverting to this everyday type of communication reveals information about the subjects practising this usage. In this sense, pragmatism becomes the theoretical basis both for formal operation and for a non-reductionistic language theory.

Manfred Eigen was correct in recognising that language and communication were and continue to be indispensable for the origin of life, the developmental processes of biological organisms, as well as for the specifically human capacity for thought, speech, and action; at the same time, he is unable to provide an adequate foundation for these two terms. This casts doubt on the entire explanatory model for living nature as provided by the biological disciplines.

1.3 The Roots of the Idea of an 'Exact' Scientific Language

Logical empiricism and critical rationalism fail in their attempt to construct a pure language of logics and mathematics as delimitation from non-scientific, metaphysical sentences. The failure is hidden in their own metaphysical concept of language upon which they cannot reflect, because they reduce the main structures of language and communication to the syntax alone. Let me reconstruct the developmental history of this misconception of language.

The origin of this can be clearly identified in the depiction theory of language of Plato. He was convinced that cosmos and the world can be sufficiently depicted by mathematics (following Pythagorean motifs). This clearly derives from his concept of ideal archetypes (the thoughts of god) which we find in this world in a variety of inexact depictions. With his language (mathematics) mankind can participate in these thoughts of god, i.e. the ideal archetypes.

Aristotle shares the view on language as a tool, but in a crucial aspect he changes Plato's conception: language acts as expression of the inner conceptions; the logical order of the linguistic sign system we use represents the logical order of nature in general. Here in the idea of an ideal language which can depict nature in a one-to-one depiction are the basics of the concept of the exact scientific language. The Aristotelian tool 'language', which functions like the tools with which we calculate mathematically, was further developed by Hobbes and Leibniz, who investigated this relationship between language and mathematics. Leibniz' intention was similar to that of his twentieth-century peers: to define a syntactic-semantic construction of our thinking in symbols and therefore to reject all misunderstandings and obscurities within sciences.

Even the concept of the young Wittgenstein postulates that behind the everyday language hides the logical form of a universal language (as postulated by Leibniz). Within this logical form of the language we can find the intersubjective and valid depiction of the fundamental facts. We have to express such fundamental facts by using these elementary sentences (*Elementarsätze*). By using these elementary sentences we can reconstruct any sensible sentences logically. The meanings of words within a language are presumed as non-variable substances which are coherent with material substances. Exactly at this point the modern empirical concept of language is congruent with the metaphysical conception of Aristotle.

This is the basics of the theory that language does not transport real contents but structures exclusively: the signs which are used within a language are variables, which have to be filled up (similar to an empty container) by communication partners from their pool of private experiences. Therefore communication is a process which starts with private encoding activities, technical transmission via a medium or communication channel and last but not least the private decoding of the receiver (sender-receiver narrative). The interpretation of certain contents being transported in the messages is purely a private matter. The unchangeable thing, the material reality, is only the logical structure of the used language.

The starting-point, again, is the Aristotelian logic of subjects and predicates which is the real depiction of the order of being. This has been picked up by scholasticism: ontology, the order of being, can be depicted in the Latin language. As noted before, according to Leibniz, this should produce a pure and logical form of speech which should lead to his programme of *characteristica universalis* independently of any meaningful content working as a universal language of sciences. (Note that here we can find Chomsky's concept of a meaning-independent, syntactical structure of universal syntax) This is exactly the purpose defined by the young Wittgenstein: to depict objective reality by one language, the language of mathematics and logic.

1.4 Postmetaphysical Thinking: Pragmatic Turn

Self-definition of the 'exact' sciences was inherently presumed to reduce every observation on this formalisable universal language. Unfortunately this failed. All attempts to translate all terms with which we express observations in terms of the theoretical language demonstrated that this was not possible in an exact way. The universal depicting language remained as a postulation that could not satisfied by real processes. Metaphysics by itself was the basis of the criticism of metaphysics by the young Wittgenstein. The depiction of the world by logical atomism (Russell's and Whitehead's *Principia mathematica*) was unmasked as secret metaphysics of logic itself. The supposition of an 'identical logical structure of language' which constitutes intersubjectivity a priori can only be simulated in computerised models in artificial binary code languages which are based on formalisable procedures. But this has nothing to do with social praxis and socially shared lifeworld of human beings. The real-life everyday language can even speak about itself; it is its own meta-language. This is not possible for identical artificially constructed languages of science, which cannot be their own meta-languages coherent with their own definition.

1.4.1 The End of Linguistic Turn

In his later work, *Philosophical Investigations*, Wittgenstein refuted the concept which he worked out earlier. The main characteristic of this pragmatic turn was the abandoning of the ideal of a world-depicting universal language. In contrast to former concepts which thought that behind any language is a material reality which determines the visible order of languages (e.g. universal laws, universal syntax) Wittgenstein proofed, that this is not the case. The most essential background of language is its concrete use in interacting humans. The real use of a language is always the unity of language and actions. This unity of language and actions Wittgenstein called *Sprachspiel* (language game). Game, because as in every game so also in language there are valid certain rules. It is not possible to go behind the practice of a life-form through explanations or foundations. Language itself is the last bastion as the real practice of actions.

Language as practical action is an intersubjective phenomenon. To insist on this fact and to demonstrate that methodological solipsism is unsuccessful in principle Wittgenstein worked out the proof of the impossibility of a private language.[11]

In his analysis of the expression 'to obey a rule', Wittgenstein provides proof that the identity of meanings logically depends on the ability to follow *intersubjectively valid rules with at least one additional* subject; there can be no *identical meanings* for the lone subject. Speaking is a form of social action. Meaning is a social function.

The rules of language games have developed historically as 'customs' from real-life usage. Such customs may even function as institutional regulations within societies. The practice of a great variety of language games is therefore the self-regulating practice of societies. They understand the rules you must play within such a game. Then you can see the meaning of a term because as co-player you get experience about how a term is used within this play, which rules determine its meaning and how the rules may change according to varying situations. Participation in common language games as precondition for the process we term 'understanding of words and sentences' is replacing the methodological-solipsistic 'empathy' by which the former concepts fill up logical structures from a private pool of experiences.

In the course of the further discussions in the philosophy of science it became increasingly clear that the validity claims of the linguistic turn could not be fulfilled (Stegmüller 1975; Apel 1976; Diederich 1978). Artificially constructed languages such as formalisable mathematical languages are totally different from natural languages such as the everyday language with which humans coordinate and organise their daily routine.

A variety of problems of formalisable scientific languages could not be solved in principle: primary as well as boundary conditions but also terms of disposition such as 'soluble', 'magnetic', 'practicability' 'progress in the cognition process' or 'visible' are not formalisable. Additionally, neither the verification criteria proposed by Carnap nor the falsification criteria proposed by Popper managed to delimitate empirical sentences from non-empirical ones (Peukert 1978). The attempt of the linguistic turn to instal logic and mathematics as the foundation of real sciences, with an unambiguously value, had to be abandoned. Linguistic turn thinkers were convinced that mathematical languages are ambiguously because they depict reality in a one to one manner. In contrast to this language even in its logical structure is an open system which cannot guarantee lack of ambiguities. The long-lasting ideal of empiricism to reduce every sentence to observation was no longer valid. Empirical theories from then on had a very risky status which only partially and indirectly can

[11] 'Is what we call 'obeying a rule' something that it would be possible for only one man to do, and to do only once in his life? (...) It is not possible that there should have been only one occasion on which someone obeyed a rule. It is not possible that there should have been only one occasion on which a report was made, an order given or understood, and so on – To obey a rule, to make a report, to give an order, to play a game of chess, are customs (uses, institutions). To understand a sentence means to understand a language. To understand a language means to be master of a technique' (Wittgenstein 1975: 80e).

be deduced by hypothesis-relaying in observations: We make observations. In the next step we make hypothesis. Out of these hypothesis we deduce conclusions. But they are not complete but partially and indirectly.

The exclusion of this history of empirical research was also a failure, as Thomas Kuhn (1967) proved. The historical set-ups and circumstances of research communities strongly influenced theory building and descriptions of observations. Progress in scientific knowledge strongly depends on social rules and group identities of scientific communities. Objectivity from then on was not an unchangeable truth but depended on consensual procedures in a great variety of language games of scientific communities.

1.4.2 The Fundamental Status of Communicative Intersubjectivity

According to these problems outlined above a theory of communicative intersubjectivity could solve these problems and therefore give a good basis for scientific rationality. This includes the withdrawal of reductionism as a formalisable term of language.

Intersubjective interactions are characterised by reciprocal validity claims. To speak, make propositions and understand utterances does not function through a private encoding process and subsequently a private decoding process, but a shared rule-governed sign-mediated reciprocal interaction. The shared competence of semiotic rules and the socialised linguistic competence to build correct sentences enable interaction partners to understand identical meanings of utterances.

The only way to decide whether a mathematical formula is true or false is by using a non-formalisable language. You cannot decide this from the formalisable language itself. With non-formalisable languages you can easily change from formalisable to non-formalisable languages and vice versa. This is impossible for the formalisable language itself. The contradictions within a formalisable language cannot be solved by this language. Therefore you need a meta-language. But some contradictions are inherent also in every meta-language. The result of this discussion was that solving these problems and paradoxes within formalisable languages and meta-languages needs a non-formalisable everyday language. This non-formalisable everyday language must be postulated as the *ultimate* meta-language. Everyday language is based on concrete social experience. A further result of this discussion was that the foundation and justification of formalisable scientific languages is possible only through a reflection on communication practice in concrete social practice of societies (Peukert 1978). Communication is a kind of social interaction, and communication science therefore has to be seen as a kind of sociology (Habermas 1984,1987). *Communicative practice of language game communities not only constitutes meanings in utterances but primarily guarantees self-identities in reciprocal interactions of common processes of social coordination and organisation.* Only the analysis of this communicative practice enables us to find essential principles of structure and function of languages in general. Even natural scientists are part of language game communities. Even the natural scientist does not start speaking and thinking just as soon as s./he starts university. Prior to this the scientist learnt linguistic and communicative competences within social interactions, as do all humans capable of language and communication.

In this discussion it became increasingly clear that every language as sign system depends on communicative agents (Böhler et al. 1986). The project to found and justify an exact scientific language failed *but it led to a highly differentiated and long-lasting reflection on language and communication which had never occurred before*. A further result of this new subject of scientific research was interest in the roots of language and communication: reflection on the inherent historicity of the interacting subjects. This means that within science this led to reflection on customs and practice of scientific communities in the light of the history of sciences (Kuhn 1967; Lakatos and Musgrave 1974). Even scientific languages are developmental processes of the practices of historically grown scientific communities. When the pragmatic turn replaced the linguistic turn this was because from now on it was not the syntax and symantics that were the central focus of investigations about languages but (i) the subjects which interact with languages as well as (ii) the pragmatic aspects in which these interacting agents are interwoven and which determine how an interactional situation is able to be constituted as such. The complementarity and non-reductionability of the three levels of rules (syntax, pragmatics, semantics) which are at the basis of any language used in communicative actions were commonsense elements (Morris 1946).

Language therefore is not solely the subject of scientific investigations of a technique for information storage or transport but depends primarily on language-using subjects with linguistic and communicative competences in real social contexts of a real lifeworld (Austin 1962; Apel 1976; Searle 1977). On the other hand, it is not possible to develop an exact language of science which functions like natural laws in inorganic matter because scientific languages are also spoken by real-life subjects and the validity claim of objectivism to eliminate all inexact parameters of subjects does not function even in the scientific language game. Also, scientific languages depend on utterances which are preliminary; they are as open as any real-life language and therefore can generate real novelties, new sentences which did not exist before, and therefore are able to progress in knowledge. Because utterances in scientific languages are subject to discourses of scientific communities and are constantly under pressure of foundation and justification they may contribute 'in the long run'(Peirce) to this progress in knowledge (Apel 1975). The meaning of words is not the result of syntactic structures solely but depends on the context within which language-using individuals are interwoven.

In the realm of this discourse on the role of language and communication in science and society the primacy of pragmatics, the level of contexts within which sign-using subjects are interwoven, became evident. The explaining-understanding controversy (Stegmüller 1975) was solved by a pragmatic communication theory which let behind the positions of classical hermeneutics and integrated speech-act theory (Wuchterl 1977; Habermas 1984, 1987, 1994). In contrast with all former concepts this pragmatic communication theory replaced the subject of knowledge of

Kant (solus ipse) by communicative intersubjective consortia of subjects that share communicative competences which enable these consortia to communicate internally as well as externally (Böhler et al. 1986). Only on this basis of communicative actions is a common understanding of identical meanings of utterances possible. This is valid also for coordination and organisation of societies.

1.4.3 Evolutionary History: History of Rule-Governed Sign-Mediated Interactions

Communication in general can be understood as rule-governed sign-mediated *interaction*. This is crucially different from chemical-physical interactions in unanimated nature, because these interactions are not governed by semiotic rules. This is equally valid for human communication and communication in non-human life.

Referring back to the rules of communicative rationality provides an opportunity to answer questions of evolutionary logic and dynamics as questions of interaction logic and dynamics (Peukert 1978). Evolutionary history can then be understood as a developmental history of interacting living agents. A more detailed examination of research results in the biological sciences should yield structures that can unequivocally be interpreted as communication rules. Understanding nature would no longer be a metaphorical expression of reductionistic explanatory models, but rather would mean understanding interaction logic and dynamics in their regulative, constitutive, and generative (innovative) dimensions.

1.4.4 Biology in the Realm of a Theory of Biocommunication

A theory of biocommunication based on a pragmatic philosophy of biology could demonstrate on the basis of empirical data that living nature in its genetic structures is language-like and in its cells, tissues, organs and organism is communicatively coordinated and organised. Karl von Frisch has proved that the interactions between honey-bees are sign-mediated, based on body behaviours which function as symbols respectively.

If this becomes the mainstream coherent description of biological processes then humans could leave their anthropocentric world view for a biocentric one, in which humans would take a new place, as being parts of a universal community of communicative living nature. This could enable biology to leave behind its mechanicism and physicalism, which are unable to differentiate clearly between life and non-living matter. Biology could start to develop as a key science with much more coherence in describing animated nature. In going back to non-reductionistic terms of 'language' and 'communication', biology could make real progress in knowledge, which would help humans in general to ensure sustainable developmental preconditions for both humans and non-human living nature. This could be a real future option for human societies in the long run.

1.5 Recent Applications of 'Language' and 'Communication' in Biology

We all noticed the entanglement of linguistics and genetics after the exploration of the universal syntax and the structural code of the DNA. Noam Chomsky's linguistic construction of a meaning-free syntax paved the way for bioinformatics and systems biology to systematise genetic content arrangements and comparative genomics. The philosophical foundation of this entanglement by Manfred Eigen failed (see above). But the entanglement of linguistics and genetics is even deeper as depiction theories can show. Let us have a look at other similar concepts.

1.5.1 Biolinguistics and Bioinformatics

Biolinguistics interprets and investigates genetic structures in the light of linguistic categories (Popov et al. 1996; Ji 1997, 1999; Searls 2002; Chomsky 2004; Zhang 2006). Similarly to bioinformatics they use statistical methods and algorithms to identify sequence orders for measurements of sequence-length and content homologies. Biolinguistics follows bioinformatics and its model of language as a quantifiable set of signs and beliefs from which it would be possible to extract semantic contents by analysis of the 'universal syntax'. In a certain sense this is possible, e.g. in genetic sequence comparison, i.e. comparative genomics. An unambiguous determination of genetic semantics through analysis of the molecular syntax of genetic code is not possible in principle, because analysis of the syntax does not tell us anything about the context in which the content bearer of the genetic information is interwoven in real life. This context plays an important role in epigenetic imprinting and therefore in the construction of different methylation patterns which then are the determinants for alternative splicing pathways of the same genetic datasets. This crucial role of pragmatic contexts is not part of the methods of biolinguistics and bioinformatics.

One result of these deficiencies is that invention of new and even complex genetic data sets or, as they may be called, gene blocks and the coherent integration of new genes or gene blocks in pre-existent genetic content arrangements by competent agents is not part of bioinformatics or biolinguistics, because innovative generation of new genetic content cannot be deduced out of a mathematic model of language, i.e. formalisable procedures such as algorithms.

Even Chomsky's attempt to reconstruct universal systems of rules within an empirical theory of language (rules that have developed over the course of evolution, are genetically transmitted, and then 'awakened' through social interaction) is founded on a 'generative grammar', which itself is based on the mathematical analysis of formal systems Chomsky (1964). He attributes the rules governing sentence construction to the level of syntax, semantics, and phonology. For him, these rules are rules of a formal system. Chomsky himself, however, concludes that formal systems are generally incapable of doing justice to the complexity of sentence structure: sentences do not appear to be produced linearly, which should be the case in formal systems. According to this model, the generating system of rules must exclude real communicative acts and interactions and, with these communicative acts and interactions, precisely the a priori of practical language usage.

1.5.2 Biosemiotics and Biohermeneutics

Biosemiotics investigates semiosis and its interpretation in living systems. Biosemiotics starts as further development of Thomas Sebeok's zoosemiotics (Sebeok 1968) and the works of Jakob von Uexküll (*Umweltlehre*) (republished by his son Thure von Uexküll 1980), who founded modern psychosomatics and human medicine on the basis of semiotic thoughts, although the term biosemiotics was used much earlier by Rothschild in the 1960s and Florkin in the 1970s. (Florkin 1974; Sebeok and Umiker-Sebeok 1992; Hoffmeyer 1996; Barbieri 2001, 2007;). Similarly to the much broader field of semiotics, biosemiotics until now has not integrated the results of the pragmatic turn and is influenced strongly by solipsistic theories of knowledge (subject-object dichotomy, message transfer within sender-receiver narratives). Parallel with this, biosemiotics is represented by diverse concepts with a natural science background such as mechanicism, physicalism, materialism, objectivism, information theory, systems theory as well as other metaphysical constructions such as ontology or even a Peirce-derived pansemioticism (everything is a sign). Most empirical biosemiotic investigations are focused on signs or the ontology of the relationship between signs or between signs and the signified something. The crucial role of pragmatics, i.e. the role of the real sign-user being part of the identity of a community of sign-users which is essential for meaning functions of signs as well as the cultural background knowledge is for interpretation processes until now has not been part of biosemiotic investigations. Currently, biosemiotics is far from being an advantageous tool for biology. If biosemiotic discourse develops certain standards in methodology this may change. Then biosemiotics could orientate biological research and interpretation of research results fundamentally.

Biohermeneutics investigates semiotic processes within and between organisms and genetic sequences such as text-like structures which can be understand hermeneutically in the realm of Gadamer and Heidegger. Interpretations of signs in sign sequences by living organisms are viewed as dialogical processes. To differentiate human communication from non-human communication by living beings the central term is enlogue as opposed to dialogue (Chebanov 1994). According to this, research in biology is proposed to proceed as hermeneutic biology, i.e. identification and interpretation of communal interacting agents and the set of signs they share. Biohermeneutics does not investigate the pragmatic rules which determine sign-mediated interactions interwoven in historically different contexts, but tries to understand semioses in living nature by (quasi)ontological hermeneutic acts.

1.5.3 Biocommunication

A first draft of a theory of biocommunication was outlined in 1975 (Tembrock 1975). Tembrock exemplified the three semiotic levels syntax, semantics and pragmatics in great detail for several behavioural patterns within the kingdom of animals. He focused on the transport of information via chemical, mechanical (tactile and acoustic) and visual signs. Although his investigations were conducted in a strict empirical manner, Tembrock justified his approach according to a solipsistic model of knowledge and communication as we came to know it in the depiction theory of language: 'There are built up inner models, of which the parameters are determined by the features of the circumstances, that are depicted by them' (Tembrock 1975, 248). His biocommunicative approach is therefore coherent with the sender-receiver model of information theory, i.e. a depiction theory similar to that of Manfred Eigen. Tembrock wants to demonstrate 'exact' science: the semioses that he investigates he tries to formalise and therefore sign-mediated interactions would be a kind of mechanistic behaviour. The inherent features of language and communication, especially the possibility of innovative semiosis or the common understanding (and interpretation) of identical meanings, is without the realm of formalisable procedures.

In contrast with this empiricist approach, at the end of the 1980s I developed a pragmatic approach of biocommunication based on the results of the philosophy of science discourse in the twentieth century (Witzany 1993, 2000, 2007). In this pragmatic conception of biocommunication I integrated the pragmatic turn in its methodological foundation as well as the complementarity of the three semiotic levels of semiotic rules. Additionally, and in contrast with theories of knowledge with a solipsistic foundation, it investigates its scientific subject according to the primacy of pragmatics, i.e. the contexts' communicative-intersubjective sign-users are interwoven in a real-life world.

The main focus of biocommunicative analysis is the *agents* that use and interpret signs in *communicative interactions*. Because 'One cannot follow rules only once' (Wittgenstein 1975), speech and communication are kinds of social behaviour and therefore it is important to investigate group behaviour and group identity, the pragmatic contexts in which they are actively interwoven together with their history and cultural identity. These groups share a repertoire of signs and semiotic rules, with which they coordinate every life organisation that is necessary.

This biocommunicative approach investigates *communicative acts* within and between cells, tissues, organs and organisms as sign-mediated interactions. The signs consist in most cases of molecules in crystallised, fluid or gaseous form and are termed semiochemicals (greek: *semeion* = sign). In higher animals additionally we can find acoustic and visual sign use. Competent sign-using agents follow syntactic, pragmatic and semantic rules in parallel. They determine the realm of possible combinations of signs, as well as interactional circumstances and the meanings of the signs within messages. No level of rules is reducible to one another. This is a crucial difference from all similar concepts of bringing together linguistics and biology.

Biocommunicative investigations concern archaea, bacteria, protoctists (eukaryotic unicellular organisms and their relatives), fungi, animals and plants. Additionally the biocommunicative approach investigates DNA/RNA sequences as code, i.e. a linguistic or language-like genetic text that underlies combinatorial (syntactic), context-sensitive (pragmatic) and content-specific (semantic) rules. From the biocommunicative perspective the interesting aspects are the linguistic (text-editing) and communicative (interaction-constituting) competences of viruses and viral-like agents such as self-replicating RNA species. The generation of meaningful nucleotide sequences and their integration into pre-existing genetic texts as well as their capability to combine, recombine and regulate these genetic texts according to context-specific (adaptational) purposes of their host organisms is of special interest in biocommunicative research.

1.6 The Structural Format of the Following Chapters

In the following chapters the descriptions of biocommunicative competences of plants, animals (bees and corals), fungi and bacteria are far from being complete. The aim was to give a representational overview of the variety of the different communicative interactions. For each aspect which is described there are a great variety of research directions and scientific articles available. Basic knowledge about biological key processes therefore is a pre-condition for reading and understanding this book.

The biocommunication processes described in this book start with the youngest of all organismic kingdoms, i.e. plants, and are followed by that of the animals in two examples: honeybees, which are of crucial importance to flowering plants, and a very old species in evolutionary terms, coral animals. After a review of the communicative competences of fungi and bacteria, the natural genome-editing competences of viruses and their important role in the evolution of life are described. Until recently their sessile non-lytic lifestyle was not mentioned or investigated very much, although their persistent lifestyle even in the DNA habitat of cellular host genomes opens a very new and interesting perspective on generation, integration, recombination and regulation of genetic text sequences. Biocommunicative aspects of telomeres and telomerases as well as the viral origins of non-coding RNAs complete this book.

The categorisation of the various levels of biocommunicative processes starts in each case with a selection of semiochemicals which serve as signs in communication processes. A next level is the interpretation processes of abiotic influences by the organisms. This is followed by the communication processes with organisms that are not related such as those from other organismic kingdoms (trans-organismic communication), which we find in a great variety of symbiotic interactions. In contrast with these, communication processes between the same and related organisms are strongly characterised by use of the same repertoire of semiochemicals and even semiotic rules (inter-organismic communication). Another level to be described is communication processes within organisms (intra-organismic communication). In the more complex organisms we can differentiate between intercellular communication and intracellular communication processes. At the end of each chapter I summarise the communicative competences of these organisms.

References

Apel KO (1975) Der Denkweg von Charles Sanders Peirce. Suhrkamp, Frankfurt

Apel KO (1976) Sprachpragmatik und Philosophie. Suhrkamp, Frankfurt

Austin JL (1962) Zur Theorie der Sprechakte. Reclam, Stuttgart

Barbieri M (2001) The Organic Codes. The Birth of Semantic Biology. PeQuod, Ancona

Barbieri M (ed) (2007) Introduction to Biosemiotics. Springer, Dordrecht

Bucher Z (1982) Natur, Materie, Kosmos. Eos, St. Ottilien

Böhler D, Nordenstam T, Skirbeckk G (1986) Die pragmatische Wende. Suhrkamp, Frankfurt

Carnap R (1931a) Überwindung der Metaphysik durch logische Analyse der Sprache. Erkenntnis 2:219–241

Carnap R (1931b) Die physikalische Sprache als Universalsprache der Wissenschaft. Erkenntnis 2:432–465

Carnap R (1934) Logische Syntax der Sprache. Wien. Schriften zur wissenschaftlichen Weltauffassung

Carnap R (1939) Foundations of Logic and Mathematics, University of Chicago Press, Chicago

Carnap R (1956) The methodological character of theoretical concepts. In: Feigl H, Scriven M (eds) Minnesota Studies in the Philosophy of Science, Vol 1, University of Minnesota Press, Minneapolis, 74–75

Carnap R (1966) Der logische Aufbau der Welt. Meiner, Hambur

Chebanov S (1994) Man as participant to natural creation. Enlogue and ideas of hermeneutics in biology. Rivista di Biologia 87:39–55

Chomsky N (1964) Current Issues in Linguistic Theory. The Hague, Mouton, London

Chomsky N (2004) Biolinguistics and the Human Capacity. Delivered at MTA, Budapest, May 17, 2004

Diederich W (1978) Theorien der Wissenschaftsgeschichte. Suhrkamp, Frankfurt

Eigen M, Winkler R (1975) Das Spiel. Naturgesetze steuern den Zufall. Pieper, München

Einstein A, Infeld L (1950) Die Evolution der Physik. Zsolnay, Wien

Florkin M (1974) Concepts of molecular biosemiotics and of molecular evolution. In: Florkin M, Stotz EH (eds) Comparative Biochemistry. Elsevier, Amsterdam

Gödel K (1931) Über formal unentscheidbare Sätze der Principia Mathematica und verwandter Systeme. Monatshefte für Mathematik und Physik 38:173–198

Habermas J (1984) The Theory of Communicative Action. Volume 1: Reason and the Rationalization of Society. Beacon Press, Boston

Habermas J (1987) The Theory of Communicative Action. Volume 2: Lifeworld and System: A Critique of Functionalist Reason. Beacon Press, Boston

Habermas J (1994) Actions, speech acts, linguistically mediated interactions and the lifeworld. Philosophical Problems Today 1:45–74

Heisenberg W (1973) Physik und Philosophie. Ullstein, Frankfurt, Berlin/Wien

Hoffmeyer J (1996) Signs of Meaning in the Universe. Indiana University Press, Bloomington

Ji S (1997) Isomorphism between cell and human languages: molecular biological, bioinformatic and linguistic implications. Biosystems 44:17–39

Ji S (1999) The linguistics of DNA: words, sentences, grammar, phonetics and semantics. Ann NY Acad Sci 870:411–417

Kuhn T (1967) Die Struktur wissenschaftlicher Revolutionen. Suhrkamp, Frankfurt

Lakatos I, Musgrave A (1974) (eds) Kritik und Erkenntnisfortschritt.Vieweg, Braunschweig

Markos A (2002) Readers of the Book of Life. Oxford University Press, Oxford, New York
Morris C (1946) Signs, Language, and Behavior. Prentice Hall, New York
Neurath O (1932) Protokollsätze. Erkenntnis 3:204–214
Peukert H (1978) Wissenschaftstheorie, Handlungstheorie, Fundamentale Theologie. Suhrkamp, Frankfurt
Popov O, Degal DM, Trifonov EN (1996) Linguistic complexity of protein sequences as compared to texts of human languages. Biosystems 38:65–74
Russel B (1940) An inquiry into Meaning and Truth. Allen and Unwin, London
Searle JR (1977) Sprechakte. Suhrkamp, Frankfurt
Searls DB (2002) The language of genes. Nature 420:211–217
Sebeok T (1968) Animal Communication. Indiany University Press, Bloomington
Sebeok T, Umiker-Sebeok J (1992) Biosemiotics: The Semiotic Web 1991. Mouton de Gruyter, Berlin
Shannon CE, Weaver W (1949) The Mathematical Theory of Communication. University of Illinois Press, Urbana
Stegmüller W (1975) Hauptströmungen der Gegenwartsphilosophie. Bd 2. Alfred Kröner Verlag, Stuttgart
Tarski A (1966) Einführung in die mathematische Logik, Vandenhoeck & Ruprecht, Göttingen
Tembrock G (1975) Biokommunikation. Vieweg, Braunschweig
Turing A (1950) Computing machinery and intelligence. Mind, 59:433–460
Uexküll T (1980) Jakob von Uexküll. Kompositionslehre der Natur. Ullstein, Frankfurt
Wittgenstein L (1975) Philosophische Untersuchungen. Suhrkamp, Frankfurt
Witzany G (1993) Natur der Sprache – Sprache der Natur. Sprachpragmatische Philosophie der Biologie. Königshausen und Neumann, Würzburg
Witzany G (2000) Life: The Communicative Structure. Libri Books on Demand, Norderstedt
Witzany G (2007) The *Logos* of the *Bios* 2. Bio-Communication. Umweb, Helsinki
Zhang HY (2006) The Evolution of genomes and language. EMBO Rep 7:748–749
Capelle W (1968) Die Vorsokratiker. Kröner, Stuttgart
Krings H (1982) Kann man die Natur verstehen? In: Kuhlmann W, Böhler D (eds) Kommunikation und Reflexion. Suhrkamp, Frankfurt
Wittgenstein L (1959) Tractatus Logico-Philosophicus. Basil Blackwell, Oxford
Wuchterl K (1977) Methoden der Gegenwartsphilosophe. Paul Haupt Verlag, Bern/Stuttgart

Chapter 2
Plant Communication

Abstract Plants are sessile, highly sensitive organisms that actively compete for environmental resources both above and below the ground. They assess their surroundings, estimate how much energy they need for particular goals, and then realise the optimum variant. They take measures to control certain environmental resources. They perceive themselves and can distinguish between 'self' and 'non-self'. This capability allows them to protect their territory. They process and evaluate information and then modify their behaviour accordingly. These highly diverse competences show us that this is possible owing to parallel communication processes in the plant body (intraorganismic), between the same and different species (interorganismic), and between plants and non-plant organisms (transorganismic). Intraorganismic communication involves sign-mediated interactions in cells (intracellular) and between cells (intercellular). Intercellular communication processes are crucial in coordinating growth and development, shape and dynamics. Such communication must function both on the local level and between widely separated plant parts. This allows plants to react in a differentiated manner to their current developmental status and physiological influences.

2.1 Introduction: Multilevel Communication Competence of Plants

Because of their apparently static life-form, plants have traditionally been viewed and treated as growth automatons. Today, however, we recognise that the coordination of growth and development in plants, as in all other organismic kingdoms, is possible only by the use of signs (Greek *semeion*) rather than pure mechanics. Understanding the use of signs in communication processes requires a differentiated perspective. Chemical molecules are used as signs. They function as signals, messenger substances, information carriers and memory media in solid, liquid or gaseous form.

G. Witzany, *Biocommunication and Natural Genome Editing*,
DOI 10.1007/978-90-481-3319-2_2,

As we will see, communicative competence refers to chemical and physical communication processes. Chemical communication is either vesicular trafficking or cell-cell communication via the plasmodesmata. Moreover, numerous signal molecules are produced in or controlled by the cell walls. Physical communication takes place through electrical, hydraulic and mechanical signs.

It should be noted that signs, whether abiotic or biotic, are interpreted. This means they must be identified as components of messages that differ from molecules that are not components of messages ('noise'). The interpreter is always a living individual. The interpretations can either be successful or unsuccessful. Thus, the message is perceived in its correct sense and meaning and a tailored response behaviour is generated, or it is misinterpreted – sense and meaning are perceived in a distorted or deformed manner – and the response behaviour fails to appear or is inappropriate.

We can recognise that the use of molecular languages/codes goes beyond information exchange: it produces various active behaviours and interactions. The many types of symbiosis show that behaviour towards the symbionts can be mutually beneficial and stress-free. Such relationships can change when this balance is lost, for example, when one partner is weakened. The interaction level shifts and one partner experiences the other as a source of stress. In plants, altruistic forms of interactions occur even in the root zone, as do life-and-death defensive battles. In every case, the situational context determines the meaning of the signs.

We will see that sign-mediated interactions within and between organisms are possible owing to the fact that living individuals share a core set of signal molecules with members of their own species and also with members of other species, families, genera or organismic kingdoms: these molecules are produced and emitted at specific levels, amounts and rhythms. The relationship between the molecules is governed by specific rules. The 'molecular syntax' (Eigen and Winkler 1975, Witzany 1995) of molecular languages/codes determines the correct sequence and combination of signal molecules. Disrupting or deforming these syntactic rules can cause incomplete transmission of the message, triggering faulty interpretations and responses in the receiver. A completely different set of rules determines the interaction behaviour between organisms, cells and tissues: growth and development are *other forms of behaviour* than defence or sexual reproduction; mutualistic symbioses require pathways that differ from those in commensalism or parasitism. One and the same core set of species-specific signal molecules are used in different interactions to produce different pathways. Moreover, one and the same pathway can take on different meanings (semantic functions) in different interactions and trigger different responses by one and the same receiver. A purely syntactic or semantic analysis cannot explain this because it cannot identify the pragmatic rules that determine the concrete interactions. This calls for consideration of all three levels of semiotic rules, as in any other analysis of sign use in living nature, and it will show the full range of multilevel communicative competences of plants.

2.2 Chemical Vocabulary of Plants

The chemical communication in and between plants is so complex that more than 20 different groups of molecules with communicatory functions have currently been identified. Up to 100,000 different substances, known as secondary metabolites, are active in the root zone, for example. This diversity is necessary given the high diversity of microbes, insects and plants in this zone (Bais et al. 2004). For example, the continuous defence against pathogenic microorganisms in the root zone requires the constant production, exact dosage and secretion of phytoalexins, defence proteins, and other substances (Flores et al. 1999). Here, I present selected examples of the molecular vocabulary in plant communication.

2.2.1 Context-Dependent Auxin as Neurotransmitter, Hormone, Morphogenic Sign

Plant roots and plant shoots detect environmental signals as well as development levels and communicate over long-distance pathways. The decentralised nervous system of plants is advantageous for decentral growth and development under constantly changing environmental conditions (Baluska et al. 2004a). Auxin is used in hormonal, morphogenic and transmitter pathways. Because the context of use can be very complex and highly diverse, identifying the momentary usage is extremely difficult (Baluska et al. 2005a). For synaptic neuronal-like cell-cell communication, plants use neurotransmitter-like auxin (Schicht et al. 2006) and presumably also neurotransmitters such as glutamate, glycine, histamine, acetylcholine, dopamine – all of which they also produce (Baluska et al. 2004a). Auxin is detected as an *extracellular* signal at the plant synapse (Baluska et al. 2005a) in order to react to light and gravity. It also serves, however, as an extracellular messenger substance to send electrical signals and functions as a synchronisation signal for cell division (Campagnoni et al. 2003). In *intracellular* signalling, auxin serves in organogenesis, cell development and differentiation. In the organogenesis of roots, for example, auxin enables cells to determine their position and their identity (Casson and Lindsey 2003). The cell wall and the organelles it contains help to regulate the signal molecules. Auxin is – as the name suggests – a growth hormone. Intracellularly, it mediates in cell division and cell elongation. At the *intercellular*, whole plant level, it supports cell division in the cambium, and at the tissue level it promotes the maturation of vascular tissue during embryonic development, organ growth and tropic responses and apical dominance (Friml and Wisniewska 2005).

2.2.2 Hormones

Alongside the classical phytohormones auxin, cytokinin, gibberellin, ethylene and abscisic acid, the plant peptide hormone systemin is observed to be important; plants use this to react systematically to local injuries (Xia 2005). For example, the abiotic stress hormone abscisic acid imparts disease resistance by acting on several levels involved in biotic stress signalling (Mauch-Mani and Mauch 2005). Peptide signal-mediated responses are merely one part of a biological process that is controlled by a combination of several hormones (Han et al. 2009). Ethylene plays a regulatory role in ethylene-sensitive flowers (Tripathi and Tuteja 2007). In activating an effective defence response, a combination of systemin, jasmonate and ethylene serves as signal molecules (Xia 2005) (Fig. 2.1).

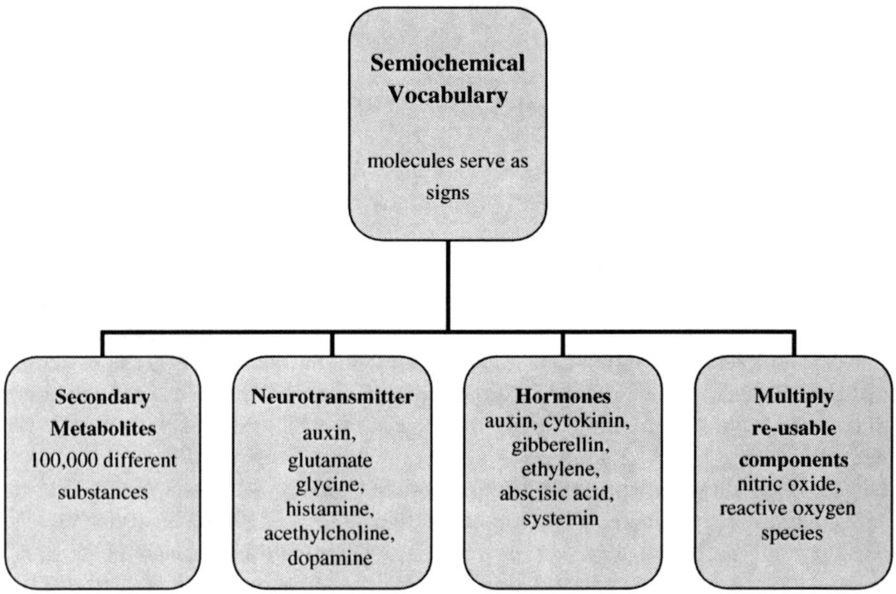

Fig. 2.1 Examples for chemical vocabulary in plant communication processes

The production (biosynthesis) of brassinolide hormones is important for cellular processes and developmental steps (Zhang et al. 2009). They are therefore termed metahormones (Amzallag 2002). Arabidopsis plants that lack this hormone remain small and are male-sterile. Many plant hormones apparently play a key role as signals in cell functions and developments that enormously impact on the activities of insects (Pearce et al. 2008). Plant hormones control not only plant growth and development but also serve in communication within the same species, with related or unrelated plant species, and with insects, i.e. they serve in classical transorganismic communication. The fact that plants and insects produce their hormones differently, but apply them for similar purposes, namely to coordinate overall development, points to their use by their unicellular ancestors (Thummel and Chory 2002).

2.2.3 RNAs

Sessile organisms can react to the full range of outside influences only through behaviours that are expressed in growth and development; correct timing, which can be very precise, is crucial (Fleming 2005). Beyond phytohormones, the chemical messenger substances include peptides such as phytosulphokine growth factors and RNAs. Micro-RNAs play an important role in intracellular communication during plant development, either in cleavage during translation/transcription or in preventing translation. Micro-RNAs are apparently necessary for meristem function, organ polarity, vascular development, floral patterning and hormone response. Many of them are developmentally or environmentally regulated (Kidner and Martienssen 2005). Small interfering RNA probably serves as a signal during early development. In later developmental phases, the RNAi-dependent epigenetic processes are reminded of this early development phase, for example, the heterochromatin configuration. In any case, these RNAs play important roles in chromatin regulation and therefore in epigenetic silencing (Kidner and Martienssen 2005).

2.2.4 Multiply Re-usable Components

Small molecules and proteins that normally support important functions in plant immunity, such as nitric oxide and Reactive Oxygen Species (ROS), have now been identified as multiply reusable components of other biological processes (Hiscock et al. 2007). Messenger substances and signal molecules are used as a versatile basic vocabulary in other contexts and other regulation networks – a common principle in the evolution, growth and development of organisms (Farmer and Schulze-Lefert 2005; Torres and Dangl 2005). Nitric oxide (NO) is a substance that has a regulatory function in numerous signal processes such as germination, growth, reproduction and disease resistance (Delledonne 2005). The same is true for diverse species of ROS (Apel and Hirt 2004; Carol and Dolan 2006).

2.3 Interpretation of Mechanical Influences

Mechanical contact has an influence on the overall organism and on the cell level, both in plants and in other eukaryotes. Contact can cause plants (a) to react aggressively, for example toward the animals that want to eat them and (b) to discard their pollen, and it can also (c) cause the plant stem to grow into the sunlight (Braam 2005). The entire configuration of a plant (morphogenesis) is partially determined by mechanical inputs, for example, wind and gravity (on the role of gravisensing: see Baluska et al. 2007; Morita and Tasaka 2004; Ross and Wolbang 2008). Responses to contact involve signal molecules and hormones along with intracellular calcium, reactive oxygen, octadecanoids and ethylenes. Another common feature is contact-related gene expression. Many of these genes code for calcium bonds, cell wall changes, defence, transcription factors and kinase proteins (Braam 2005).

The detection of resources and their periodic, cyclic availability plays a key role in plant memory, planning, growth and development. When, for example, young trees obtain water only once a year, they learn to adjust to this over the following years and concentrate their entire growth and development precisely in the expected period (Hellmeier et al. 1997).

Interpretation processes in the plant body are highly sensitive. In taller-growing plants, for example, the water balance places enormous demands on cell wall development and cell wall structures, which must adapt to the (often extreme) pressures involved in storage and pressure distribution. A sophisticated and multi-levelled feedback – and feedforward – system guarantees a plant-compatible water balance even under extreme environmental conditions (Zimmermann et al. 2004; Buckley 2005). To date, seven different levels of sensitivity to water shortage have been described. They are based on the different types of physiological and phenotypic responses (Trewavas 2005).

Plants are especially sensitive to light and have various receptors for UV, blue, green, red and far-red light (Trewavas 2005). The angle of the light, combined with the sensation of the growth of adjoining plants, is decisive in enabling plants to coordinate their growth with respect to the optimal light angle and shade avoidance (Ballare 1999). The adaptive response of the plant, i.e. altered growth, depends on the seconds-, minutes- and hours-long dominating wavelength of the incoming light, and on the combination of wavelengths across the whole day. The roots receive constant signals from the aboveground parts of the plant for specific growth orientations (Baluska et al. 2006).

2.4 Transorganismic (Transspecific) Communication

Sign-mediated interactions with organisms belonging to other species, genera, families and organismic kingdoms are vital for plants and are coordinated and organised in parallel. They are almost always symbiotic or parasitic and range from mutually beneficial via neutral to damaging behaviours. The different forms of symbiotic communication require very different behaviours from the participating partners. This involves large numbers of complementary direct and indirect defence behaviours.

2.4.1 Coordination of Defence against Pests and Injury

A good example of parallel trans-, inter- and intraorganismic communication is the coordinated defence strategies of plants. Chemical signal substances are the oldest form of signs and are used by microbes, fungi, animals and plants. They are transmitted via liquids in the environment or within the plant body; they can be distributed and perceived through the atmosphere. Leaves always emit such volatiles in small doses, but emit greater quantities when infested by parasitic insects. This allows them to attack the parasites either directly by producing substances that deter them, or indirectly by attracting other insects that are natural enemies of the parasites. These volatiles are also perceived by neighbouring plants, allowing them to initiate pre-emptive defensive responses (Paré and Tumlinson 1999). Volatile phytochemicals serve as airborne *semiochemicals*. Depending on the behavioural context – destruction, injury or parasitic infestation – the emitted scents clearly differ for both the insects and neighbouring plants (Paré and Tumlinson 1999). The plants coordinate complementary direct and indirect defence mechanisms in a stepwise manner and tailor them flexibly to the severity of the injury or the density of pest infestation (Kant et al. 2004; Engelberth et al. 2004; Wen et al. 2007).

When plants are attacked by pests, they develop immune substances that function the same way as in animals (Nürnberger et al. 2004). Injured plants produce aromatic substances that warn other plants. They then rapidly produce enzymes that make the leaves unpalatable for herbivorous insects. Rather than being passive 'prisoners' of their surroundings, plants are active organisms (Peak et al. 2004) that identify their pests and actively promote the enemies of these pests (van der Putten et al. 2001).

In lima beans, for example, a total of five different defence strategies against mite infestation have been discovered. First, they change their scent to make themselves unattractive to the mites. Then the plants emit scents that are perceived by other plants, which then do precisely the same thing to warn surrounding lima beans before the mites even reach them. Some of the emitted substances have the effect of attracting other mites that eat the attacking red mites (Mithöfer et al. 2005). Similar defence processes have been described in tomato plants (Pearce and Ryan 2003; Kant et al. 2004).

Plants possess a 'non-self' warning system to fend off dangerous parasites. So-called pattern recognition receptors detect patterns of chemical substances associated with parasite infestation (Zipfel and Felix 2005). The microbes, in turn, react to this pattern recognition (Nomura et al. 2005).

Because plants are sessile, their reaction potential is geared toward defence against mechanical damage and pest infestation (De Vos et al. 2007). One of the many reaction types to infestation is the production of protease inhibitors I and II, which block protein degradation in the digestive tracts of insects. This defence reaction is produced at both the injured site and throughout the surrounding tissue: the local wound response triggers the production of mobile signals that prompt a systematic reaction of the overall plant (Xia 2005).

Plant roots have the capacity to produce 100,000 different compounds, largely secondary metabolites, many with cytotoxic properties, in order to prevent the spread of microbes, insects and other plants (Bais et al. 2004, Walker et al. 2003) For example, plants have developed defence strategies in which substances are emitted in the root zone such as signal mimics, signal blockers and/or signal-degrading enzymes to respond to bacterial quorum-sensing (Walker et al. 2003). In the defensive position, they can disrupt the communication of parasitic microorganisms to the point that the internal coordination of the parasitic behaviour collapses.

'Friendly' arthropods such as predaceous or fungivorous mites are supported by plant 'domatia', similarly to the situation in complex communities of grasses and fungal endophytes. These symbiospheres, however, can also be misused, for example, by mites that colonise these domatia for themselves without benefiting the host cell (Romero and Benson 2005).

2.4.2 Communicative Coordination of Symbioses

A limited number of chemical messenger substances is available to maintain and simultaneously conduct the communication between (a) root cells of three different types, (b) root cells and microorganisms, (c) root cells and fungi, and (d) root cells and insects (Bais et al. 2004, Callaway 2002; Dessaux 2004; Dunn and Handelsman 2002; Teplitski et al. 2000; Walker et al. 2003). The communication process in the root zone is generally trans-, inter- and intraorganismic and requires a high communicative competence in order to be successfully interactive on all three levels and to distinguish messenger molecules from 'noise' (Federle and Bassler 2003; Hirsch et al. 2003; Sharma et al. 2003). It has been postulated that the origin of root cells in plants, and therefore the basis for the youngest organismic kingdom on our planet, arose through the symbiogenesis of fungi and algae (Jorgensen 1993; Zyalalov 2004; Baluska et al. 2006). One hypothesis assumes that land plants are the symbiogenetic product of green algae and a tip-growing fungus-like organism that combined autotrophic and heterotrophic capabilities (Jorgensen 2004).

2.4.3 Vital Symbiosis of Plant Roots with Bacteria, Fungi and Animals

Plants use their plant-specific synapses (Baluska et al. 2005a) to conduct neuronal-like activities and establish symbiotic relationships with bacteria (Denison and Kiers 2004). Similar mutually advantageous relationships are established with mycorrhizal fungi (Vandenkoornhuyse et al. 2002). A special type of plant synapse resembles the immunological synapse of animal cells and allows plants to respond to pathogen and parasite attacks as well as to establish stable symbiotic interactions with rhizobia bacteria and fungal mycorrhiza (Baluska et al. 2005a; *see also*: Estabrock and Yoder 1998; Yoder 1999; Keyes et al. 2000; Kahmann and Basse 2001; Engelbert et al. 2004; Imaizumi-Anraku et al. 2005). Electrical signals can reinforce chemical signals or overcome short-distance responses of fungal mycelia that can be present on root surfaces (Pieter van West et al. 2002). Interestingly, rhizobia bacteria are taken up in plant cells via phagocytosis during symbiotic interactions with roots of leguminous plants (Samaj et al. 2004). The symbiotic relationship between legumes and rhizobial bacteria leads to the formation of nitrogen-binding nodules in the root zone (Lee and Hirsch 2006). Nod factor signalling and thigmotrophic responses of root hairs overlap here as well. This once

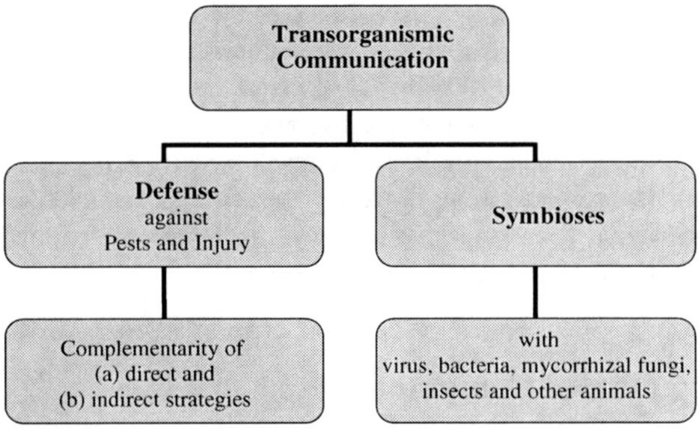

Fig. 2.2 Examples for transorganismic (transspecific) communication

again shows how the same pathways are used for different signal processes (Guerts et al. 2005) (Fig. 2.2).

Today, several hundred species of fungi colonise more than 100,000 different plant species. This type of cohabitation requires symbiotic signalling (Lammers 2004). Roots develop from rhizomes in order to provide better conditions for mycorrhizal fungi, which in turn supply plants with better nutrients (Brundrett 2002). For the fungus, the relationship is either balanced or predatory. Endophytic fungi, however, live in plants without triggering disease symptoms (Brundrett 2002). Similarly to the symbiosis between plants and mycorrhizal fungi, the symbiosis between asexual endophytes and grasses also represents a type of complementary parasitism (Müller and Kraus 2005).

Plants, insects and microbes share a particular repertoire of signals (Kempema et al. 2007). Some are therefore also employed strategically. Thus, plants also use insect hormones (prostaglandins) for specific defence behaviour. Signal theft is common. Because plants can detect their own signals, they can presumably also detect similar signals that are used in communication between insects (Schultz and Appel 2004).

2.4.4 Viral Symbiotic Interactions

In particular, the evolution of plant viruses shows that viruses complement plants both competitively and symbiotically. A healthy plant body is better for most viruses than a sick body. Plant viruses and their development provide a good explanation for the observation that new species originate through symbiogenesis (Roossinck 2005). Viruses use intergenomic gene transfer and intragenomic duplication. Many DNA viruses have encoded numerous nucleic acid metabolisms that are very similar to cell proteins. Examples include DNA polymerases, ribonucleotide reductase subunits, DNA-dependent RNA polymerase II subunits, DNA topoisomerase II, thymidylate synthase, helicases and exorbinuclease. Viruses probably invented DNA to protect their genetic material from being changed by RNA or RNA-encoded enzymes (Shackelton and Holmes 2004). One of the interaction processes between plant viruses and their host organisms creates a defence level against foreign nucleic acids (Dunoyer and Voinnet 2005a). Plant viruses code for silencing suppressors in order to act against host RNA silencing, and some of these suppressors effect micro-RNA multiplication and hinder plant development (Wang and Metzlaff 2005). Viroids also play a symbiotic role, however. Despite their small size and their non-coded genome, viroids can multiply, systematically spread from cell to cell, and trigger symptoms in the host (Dunoyer and Voinnet 2005b).

2.5 Interorganismic Communication

Research has shown that plants can distinguish between damage caused by insects and mechanical injuries. Mechanically-injured plants emitted substances such as volatiles that were ignored by neighbouring plants, whereas they all reacted immediately to pest infestation (Mescher et al. 2006).

Plants can distinguish between 'self' and 'non-self'. Thus, defence activities are initiated against foreign roots in order to protect the plant's own root zone against intruders. The individual sphere of a root, along with its symbiotic partners, requires certain fundamental conditions in order to survive and thrive. When these prerequisites are threatened by the roots of other plants, substances are produced and released in the root zone that hinder this advance (Bais et al. 2003; Dunn and Handelsmann 2002, Dessaux 2004, Walker et al. 2003). Such defence activities are also deployed as anti-microbial substances against the microflora in the root zone (Fig. 2.3).

Plant roots produce a wide range of chemical substances: (a) some enable species-specific interactions; (b) many of these substances are released tens of centimetres into the surroundings; (c) these substances have strong but not necessarily negative effects on animals, bacteria, viruses and fungi; (d) released substances

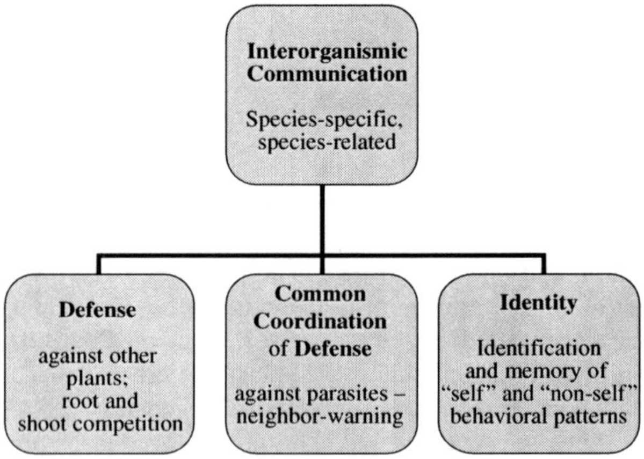

Fig. 2.3 Examples for interorganismic (species-specific, species-related) communication

have a defensive function against other plants; (e) many substances have absorptive characteristics that reduce the negative effects of substances (Bais et al. 2003).

As reported above for lima beans and tomatoes, corn plants also use a sophisticated communication system to warn each other about pests. By emitting green leafy volatiles, the corn plants attract the natural enemies of the pests and alarm neighbouring plants. The alarmed neighbour then produces a protective acid that is normally produced only in response to external injuries (Engelberth et al. 2004). Plants use biotic signals to inform each other about the presence, absence and identity of neighbouring plants, growth space, growth disturbances and competition (Callaway et al. 2002). Plants that are removed and planted elsewhere remember the identity of their former closest neighbours for several months (Turkington et al. 1991). Recognition patterns in neuronal-like networks are one possible explanation.

Parasitic plants are an important feature in the plant world. Today, about 4000 species have been described. In order to parasite other plants, their root apices transform into fungal-like haustoria which extract photosynthates from vascular tissue of prey roots (Yoder 1999; Tomilov et al. 2005). Parasitic plants are present wherever other plants can grow, from the tropical rainforest to the Arctic, and take important nutrients and environmental resources (light) away from non-parasitic plants. They therefore influence entire ecosystems, population dynamics, and biodiversity, including the presence and diversity of microbes, birds, insects and other animals (Press and Phoenix 2005).

2.6 Intraorganismic Communication

Unlike the central nervous system of animals, which controls metabolism and reactions centrally, the control in plants is decentral (La Cerra and Bingham 2002). This enables plants to start independent growth or developmental activities in certain regions of their body, for example, deciding on how a particular branch should grow, depending on the wind, light angle and overall 'architecture' of the plant (Trewavas 2005). Most of the activities in which plants engage with regard to growth and development require communication processes – synapse-like communication – between all parts of the plant.

2.6.1 Most Intercellular Communication via Plasmodesmata

Short-distance communication differs considerably from long-distance communication. As a rule, they complement each other. Intercellular communication in the root zone (in the soil) differs from that in the stem region above ground. Both are necessarily coordinated with one another in order to enable life in these different habitats. Intercellular communication informs other plant parts about events in specific organs or regions of the plant (especially in large plants), for example, sugar production in leaves, reproduction in flowers and resource utilisation by roots (Xoconostle-Cázares et al. 1999).

Plant cells are connected by plasmodesmata. These connecting channels enable the flow of small molecules as well as ions, metabolites and hormones, and allow the selective exchange (size exclusion limit) of macromolecules such as proteins, RNAs and even cell bodies (Baluska et al. 2004b). The plasmodesmata endow plants with a cytoplasmatic continuum known as the symplasm (Dunoyer and Voinnet 2005b). Plasmodesmata, however, are more than mere transport channels; they also regulate and control the exchange of messenger substances in a very complex manner (Gillespie and Oparka 2005). In symplastic signalling, the intercellular communication of plants differs fundamentally from that in other organismic kingdoms (Golz 2005). It integrates various communication types such as local and long-distance communication. Beyond symplastic communication (especially in the meristem, where new tissues are produced), plants also exhibit the receptor-ligand communication typical of animals (Golz 2005). While receptor-ligand communication determines stomatal patterning in the epidermis of mature leaves, trichome patterning is mediated by symplastic signalling (Srinivas and Hülskamp 2005).

For long-distance signalling movement proteins play an important role. Movement proteins convey information-bearing RNA from the stem and leaves to the remote roots and flowers. The movement protein allows the mRNA to enter the plasmodesmata tunnel into the phloem flow. Once it has entered this transport system, it can relatively rapidly reach all parts of the plant. These RNAs can control the levels of other proteins. The level contains information for local tissues, for example, about the general physical condition of the plant, the season, or the presence of dangerous enemies (Xoconostle-Cázares et al. 1999).

Plasmodesmata are prerequisites for intercellular communication in higher plants (Tassetto et al. 2005). In embryogenesis they are an important information channel between foetal and maternal tissue. The greater the development of the embryo, the more reduced the cell-cell communication between embryo and maternal tissue (Kim and Zambryski 2005). Cell-cell communication via direct transmission of transcription factors plays a central role in root radial and epidermal cell patterning as well as in shoot organogenesis (Kurata et al. 2005). The cellular organisation of the roots is determined during the plant's embryonic development and is controlled by intercellular communication. Bonke and colleagues (2005) provide a particularly good example of communicative control of these ten phases of embryogenesis. This confirms the presence of local signalling centres and the complex relationship between numerous different signalling pathways.

A wounded plant organises an integrated molecular, biochemical and cell biological response. This strategy enables information to be transported across great distances, for example, in tall trees (Schilmiller and Howe 2005). Proteins that can be detected by receptors enable a 'thoughtful response' (McClintock 1984) by plants. There are about one thousand known protein kinases/phosphatases, numerous secondary messengers and many thousands of other proteins (Trewavas 2005). Throughout their lifecycles and their growth zones, plants develop a 'life history' of environmental experience that they can pass on to later generations and, should they grow to be several hundred years old, utilise themselves (Trewavas 2005). Even small plants store stress experiences in their memories and then use these memories to coordinate future activities (Goh et al. 2003). Especially during growth, key information about the current status often takes a back seat to future-oriented processes, for example, early root growth and nutrient supply to secure future developments such as larger leaves. From this perspective, plants must plan for the future and coordinate growth, food uptake and communication with symbionts (Trewavas 2003a,b).

The complementary differentiation of communication types into short-distance and long-distance signalling – with their different yet ultimately complementary tasks – requires cells to identify their position. They accomplish this by, among other things, detecting signals from neighbouring cells (Coupland 2005). Thus, the identification competence of 'self' and 'non-self' by cells can be interpreted as a result of social interaction rather than solipsistic behaviour. For example, signals from leaves trigger flower development at the tip of a plant (Coupland 2005). An entire network involving four different signal pathways regulates this transition from the vegetative to the reproductive phase (Coupland 2005). Most flowers bear closely adjoining male and female reproductive organs. Self-incompatibility is therefore crucial in distinguishing between own (related) and foreign (non-related) pollen. This self/non-self differentiation ability is promoted by signal processes also used in other plant responses (McCubbin 2005).

Signals amend one another to form signal sequences, much like words combine to form sentences: different active forms of behaviour determine the combination and production of molecule 'sentences'. This distinguishes cell differentiation during root development from cell differentiation during stem development, or developmental processes during the vegetative phase from developmental processes in the reproductive phase.

2.6.2 Intracellular Communication

Intracellular communication in plants takes place between the symbiogenetically assimilated unicellular ancestors of the eukaryotic cell, mainly between the cell body and cell periphery. It transforms and transmits external messages into internal messages that exert a direct (epigenetic) influence on the DNA storage medium and trigger genetic processes; this leads to the production of signal molecules that generate a response behaviour. Via endocytosis, however, bacteria, viruses and viroids interfere with this intracellular communication and can support, disrupt or even destroy it. Intracellular communication offers viruses the opportunity to integrate certain genetically-coded abilities of the host into their own genome or to integrate their own genetic datasets into the host genome. The ability of viruses to integrate different genetic datasets probably plays a major role in symbiogenetic processes. The eukaryotic cell is composed of a multicompetent nucleus as a basic building block of life and a cell periphery 'apparatus' that was symbiogenetically the ancestor of other endosymbionts. Interestingly, both nucleus and viruses have several similar features and capabilities: they both lack the protein synthesis 'machinery' and the fatty acid-producing pathways. Both transcribe DNA but do not translate it into RNA. Viruses were probably very important in the evolution of eukaryotic cells because they were able to conduct cell-cell 'fusion' (Baluska et al. 2005b). There are strong reasons, too, why the eukaryotic nucleus is of viral origin (Bell 2001; Ryan 2002; Villarreal 2005) (Fig. 2.4).

Neuronal plasticity refers to the ability of neuron populations to alter – either to strengthen or weaken – their connections *based on experience*. This is the basis for learning and memory. Like memory, long-term neuronal plasticity requires new RNA and protein synthesis. Accordingly, the signals must be transported from the

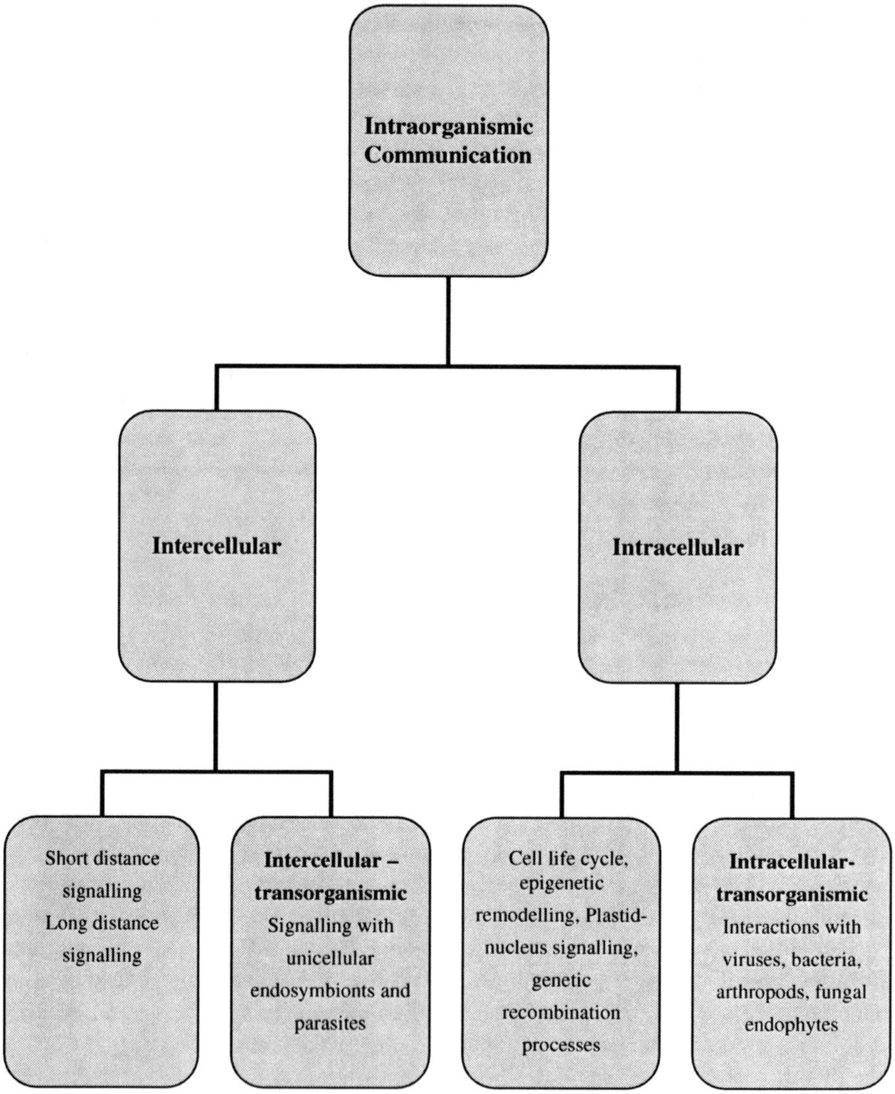

Fig. 2.4 Selected examples for intraorganismic communication

synapse, from where they are sent, to the nucleus, where they are transformed to change the gene transcription. Then, the products of gene transcription (proteins, RNAs) must be sent back to the synapse in order to permanently change synaptic strength. This communication process is well described in animals (Thompson et al. 2004, Martin 2004, Moccia et al. 2003); if plants exhibit neuronal plasticity, then similar descriptions may follow.

Reports on the transfer of mitochondrial genes between unrelated plant species caused some surprise. While gene transfer is an extremely rare event in animals and fungi, it is common between plant mitochondria (Andersson 2005). Variations in repetitive DNA that manifest themselves as variations in the nuclear DNA complex have far-reaching ecological and life-history consequences for plants (Meagher and Vassiliadis 2005).

The function of a eukaryotic cell depends on successful communication between its various parts. Plastids send signals to regulate nuclear gene expression and thus to reorganise macromolecules in response to environmental influences (Strand 2004). It has been shown that micro-RNAs regulate certain developmental processes such as organ separation, polarity and identity, and that they define their own biogenesis and function (Dugas and Bartel 2004). Eukaryotic genomes are regionally divided into transcriptionally active euchromatin and transcriptionally inactive heterochromatin (Bender 2004). Epigenetic changes can also take place without changes in genomes, for example, through various inactivations and activations of genetic datasets via chromatin remodelling, transposon/retro release, DNA methylation, novel transcription, histone modification, and transcription factor interactions (Jablonka and Lamb 2002). Epigenetic changes are also reversible (Rapp and Wendel 2005). Various stress situations in plants are known to cause transposon movements (Kumar and Bennetzen 1999), and bacterial infections or UV stress can cause chromosomal rearrangements (Kovalchuk et al. 2003), i.e. changes in higher-order regulation levels that control the transcription processes of the protein-coding DNA. Also the defence activation of innate immunity in the case of microbiological infections depends on signalling processes (Ma et al. 2008).

Repetitive DNA is present in two syntactic combinations: tandem repeats and dispersed repeats. Tandem repeats consist of sequences that can contain several thousand copies of elements that are dispersed throughout the genome. Pericentromeric sequences consist of a central repetitive nucleus flanked by moderately repetitive DNA. Telomeric and subtelomeric sequences consist of tandem repeats at the physical end of the chromosomes. Retroelements and transposable elements are involved in replication and reinsertion at various sites in complex processes: these include activation of excision, DNA-dependent RNA transcription, translation of RNA into functioning proteins, RNA-dependent DNA synthesis (reverse transcription) and reintegration of newly-produced retroelement copies into the genome (Meagher and Vassiliadis 2005).

Endocytosis and vesicle recycling via secretory endosomes are indispensable for many processes in multicellular organisms. Plant endocytosis and endosomes are important for auxin-mediated cell-cell communication as well as for gravitropic responses, stomatal movements, cytokinesis and cell wall morphogenesis (Samaj et al. 2005). As in animals, synaptic cell-cell communication is based on rapid endocytosis and vesicular recycling in plants (Samaj et al. 2005).

Plants can overwrite the genetic code they inherited from their parents and revert to that of their grand- or great-grandparents (Lolle et al. 2005; Weigel and Juergens 2005; Pearson 2005). This contradicts traditional DNA-textbook conviction that children simply receive combinations of the genes carried by their parents. Recently a backup code has been found; it can bypass unhealthy sequences inherited from the parents and revert to the healthier sequences borne by their grandparents or great-grandparents. Research has shown that plants are able to replace abnormal parental code sequences with the regular code possessed by earlier generations. Does this require inheritance not only of the parental genetic make-up but also that of the grandparents and former ancestors? What is proposed is that higher-order regulation function in non-coding DNA saves ancestor genome structures, which overrule protein-coding DNA under certain circumstances like stress (Witzany 2005). This means that the (pragmatic) situational context of the living plant body may induce epigenetic intervention, i.e. active micro-RNAs activate a certain signalling pathway network which can restructure the semantics of a genetic make-up. By initiating chromosomal methylation and histone-modifications, certain silencings, start and stops, and alternative splicing processes constitute alternative sequences. The result is that, in the existing genome architecture, it is not the inherited parental sequences that are translated and transcribed but the backup copy of grand- or great- grandparents. Under normal conditions, the operative genetic make-up stems from the parents. These research results indicate that not only is a combination of parental genes inherited, but also ancestral genome-regulating features in 'non-coding' DNA; this enables alternative splicing pathways, i.e. a different use and multiple protein meanings of one and the same genetic data set (Lolle et al. 2005; Weigel and Juergens 2005; Pearson 2005).

2.7 Plant Communication: Plant Neurobiology and the Emergence of Mind?

Since new detections on the synapse-like communication in plants have become relevant, plants seem to assemble a variety of behavioural patterns that are a kind of intelligent behaviour such as constant measurements of the surroundings, memory building, learning, fine-tuned emission of volatiles to warn neighbouring plants, parallel communication in the root zone with mychorizal fungi, rhizobacteria, insects, and even mimicry. The use of the term 'intelligence' caused a controversy (Trewavas 2003a, 2003b, 2005) which led to a further controversy on plant neurobiology also (Brenner et al. 2006, 2007; Alpi et al. 2007; Trewavas 2007; Barlow 2008; Baluska and Mancuso 2007). To the proponents of traditional textbook conviction these terms are pure metaphors which cannot be substantiated in reality. To the researchers who focus on neuronal synapse-like biocommunication they have an essential background and can be substantiated.

If we look at this controversy from a biocommunicative perspective we may be able to solve it by stating the following: We may be able to contribute following aspects:

- The discussion as it occurred is a necessary part of every real developmental process in science. It includes philosophical aims (mechanistic vs. organismic), psychological aims (conservative vs. progressive) anthropological aims (human cognition apparatus) and sociological aims (group identity of research teams). To find an appropriate way of use is therefore not easy because biology has no method for answering philosophical questions and very restricted possibilities for handling psychological, anthropological and sociological problems. Nor can we neglect the role of history-dependent influences on theory building, as outlined with great success by Thomas Kuhn (1962).
- The capabilities of plants for learning, memory, self-recognition, even a kind of decision-making, are strong indicators of complex organismic behaviour. Certainly, it is also a kind of intelligent behaviour. But this is not astonishing. Also the investigations on the life of bacteria show clear examples of social intelligence: 'Bacteria have developed intricate communication capabilities (e.g. quorum-sensing, chemotactic signalling and plasmid exchange) to cooperatively self-organize into highly structured colonies with elevated environmental adaptability. We propose that bacteria use their intracellular flexibility, involving signal transduction networks and genomic plasticity, to collectively maintain linguistic communication: self and shared interpretations of chemical cues, exchange of chemical messages (semantic) and dialogues (pragmatic). Meaning-based communication permits colonial identity, intentional behavior (e.g. pheromone-based courtship for mating), purposeful alteration of colony structure (e.g. formation of fruiting bodies), decision-making (e.g. to sporulate) and the recognition and identification of other colonies – features we might begin to associate with a bacterial social intelligence. Such a social intelligence, should it exist, would require going beyond communication to encompass unknown additional intracellular processes to generate inheritable colonial memory and commonly shared genomic context' (Ben Jacob 2004).
- So the problem is not if plants have intelligence, mind, consciousness or even altruistic behaviour. The problem is how we can define such terms; are they mechanistic, empiristic, ontological, metaphysical, objectivistic, naturalistic, progressive, conservative or even action theoretical?
- Humans working as scientists cannot exclude the fact that their cognition apparatus has its roots in typical features of central nervous system capabilities. That is why it is so interesting to observe and describe behavioural patterns of similar constructed organisms, like mammals. To observe and describe an exclusive sessile lifestyle like plants with very different behavioural sequences determined by a logic of coordination which depends on the capabilities of the opposite, a decentralised organisation, is difficult per se, although *behavioural action patterns* are quite similar to other organisms (mating, defence, attack, nutrition sensing, etc.)
- Whether it is justifiable to use plant neurobiology does not depend on whether plant neuronal-like processes are very similar to those of animals, but whether

they use chemical pathways that are similar to neuronal communication in general. This means neuronal communication in plants can also have a very different structure in timescale from that of animals. Nevertheless if the key features are similar it seems to be justifiable to speak about plant neurobiology even if this is unfamiliar to conservative scientists, because it can serve as an explanatory pattern which can describe phenomena from a more appropriate perspective.
- To escape metaphysical mousetraps, I would suggest not using terms like mind or consciousness. Both terms even today are far from a coherent non-reductive definition even in humans. Conversely, undoubtedly coherent is our definition from communication processes as sign-mediated interactions coherent to syntactic, pragmatic and semantic rules. This means that independently of whether some assume plants to have mind or consciousness plants indeed have a great variety of communicative competences which are of similar complexity to those of animals, although in quite different processual pathways. Without these communicative competences the coordination of complex organisation of plant life could not happen. Parts of these communicative competences show clear neuronal-like signalling pathways.

2.8 Conclusion

Plants are the youngest organismic kingdom and perhaps the main success story of evolution. They arose ca. 350 million years ago, and terrestrial plants, which flower and bear fruits (a key prerequisite for feeding in larger animals), only developed 150 million years ago. Higher plants make up 99% of the biomass on our planet; of this, nearly 84% are trees. The lack of mobility is often construed as a disadvantage vis-à-vis representatives of the animal kingdom. From an objective perspective, such immobility and the sessile lifestyle must have been an advantage. Plants are clearly the most malleable of organisms, a trait that can be attributed to the symbiogenetic unification of five to seven different unicellular, ancestral organisms.

An ever-increasing body of data shows that evolution, growth and development as in all other organismic kingdoms – depend on successful communication processes. This is the prerequisite for the internal coordination and organisation of the organism and its interplay with other organisms. Communication processes, however, go beyond mere information exchange to include highly differentiated and manifold sign-mediated interactions. The signals used in these interaction processes underlie certain semiotic rules. The rule adherence is very reliable and conservative. Nonetheless, the rules governing sign use can be damaged, incompletely executed, deformed, abandoned, or even newly-generated. Therefore biosemiotic rules – as opposed to natural laws – are principally changeable (Fig. 2.5).

As in all sign processes (semioses) in living nature, *syntactic* rules determine the relationship of the signs to one another, i.e. provide combination rules for the sequence, combinatory ability, density and rhythm of the signs used. The syntactic rules differ from *pragmatic* rules. They enable entirely different interactions in that the interaction partners must generate a very specific behaviour in order for

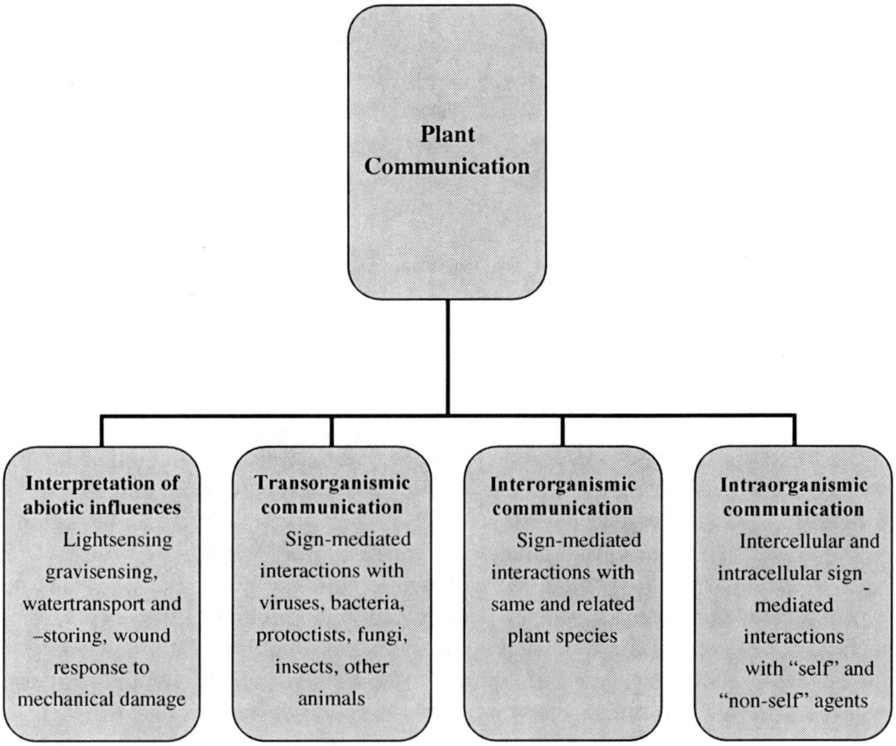

Fig. 2.5 Communication processes are sign-mediated interactions which each obey three levels of semiotic rules (syntactic, pragmatic, semantic)

interactions, for example, mutual coordination and organisation, to be successful at all. Here, concrete life situations with very specific behavioural contexts are involved. Depending on the context of use, one and the same syntactically correct sign sequence of chemical signal molecules can take on different meanings (*semantic* functions).

Plants fundamentally depend on successful communication. The behaviour in the specific interaction can be misinterpreted. A plant can feign mutualism, for example, in order to gain a one-sided advantage from the interaction and to damage, permanently exploit or kill the partner. This, however, cannot be the representative form of communication because no individuals would survive if all plants behaved in this manner. The majority of interactions must be successful for several participants.

Communication processes are successful when the rules governing sign use are correctly followed. Clearly, rules can be broken. In such cases, the messages transmitted via the signs are incomplete, incorrect, and induce no or an inappropriate behavioural response. Messages can also be misinterpreted: The sign user uses (a) the sign incorrectly/misleadingly, and the message does not arrive because it is mutilated, fragmentary, and in due course, the recipient cannot respond to the message in the manner required by the non-mutilated message. (b) The sign continuously expresses a message that does not conform to reality ('insect enemies are attacking'); the recipient of the message will respond in a manner adapted to the reality of this nonconformity. (c) The message is used to mean something other than it is normally used to convey (in order to gain one-sided advantages). Any constant rule-breaking blocks the organisation of life processes (communicative coordination of evolution, reproduction, growth, development) within and between organisms.

The term semiochemicals was generally used to designate molecules that served in communication between organisms. As the present review demonstrates, all chemicals which function as signs in sign-mediated interactions in and between organisms are semiochemicals. This would be compatible with the biosemiotic approach, which considers the full range of sign use within and between living organisms (Witzany 2006).

In the future, the trans-, inter- and intraorganismic levels of communication processes will be better understood. This, in turn, will allow better differentiation of the different levels of rules that govern signal use. The syntax of intracellular sign use differs from the syntax at the intercellular level, as well as from the syntax of the signs in species-specific interactions, which in turn differs from sign use in transspecific interactions between organisms. Embedded in the ecological framework, these rules for constituting sign-mediated interactions are used differently depending on the behavioural context. One and the same message can contain, in other contexts, entirely different meanings. Integrating this biosemiotic perspective will help us more gradually to decipher the specific meaning of the full range of semiochemicals (in their broader sense) and to be aware of the high-level communicative competences of plants.

References

Alpi A, Amrhein N, Bertl A et al. (2007) Plant neurobiology: no brain, no gain? Trends Plant Sci 12:135–136

Amzallag GN (2002) Brassinosteroids as metahormones: evidences from specific influence during critical period in Sorghum develpoment. Plant Biol 4:656–663

Andersson J (2005) Lateral gene transfer in eukaryotes. Cell Mol Life Sci 62:1182–1197

Apel K, Hirt H (2004) Reactive oxygen species: metabolism, oxidative stress, and signal transduction. Ann Rev Plant Biol 55:373–399

Bais HP, Park SW, Weir TL et al. (2003) How plants communicate using the underground information superhighway. Trends Plant Sci 9:26–32

Bais HP, Park SW, Weir TL, Callaway RM, Vivanco JM (2004) How plants communicate using the underground information superhighway. Trends Plant Sci 9:26–32

Ballare CL (1999) Keeping up with the neighbours: phytochrome sensing and other signalling mechanisms. Trends Plant Sci 4:97–102

Baluska F, Mancuso S, Volkmann D et al. (2004) Root apices as plant command centres: the unique, 'brain-like' status of the root apex transition zone. Biologia (Bratislava) 59:7–19

Baluska F, Volkmann D, Menzel D (2005) Plant synapses: actin-based domains for cell-to-cell communication. Trends in Plant Sci 10:106–111

Baluska F, Volkman D, Hlavacka A, Mancuso S et al. (2006) Neurobiological view of plants and their body plan. In: Baluska F, Mancuso S, Volkmann D (eds) Communication in Plants. Springer, Berlin/Heidelberg

Baluska F, Barlow PW, Volkmann D et al. (2007) Gravity related paradoxes in plants: plant neurobiology provides the means for their resolution. In: Witzany G (ed) Biosemiotics in Transdisciplinary Contexts. Umweb, Helsinki

Baluska F, Mancuso S (2007) Plant neurobiology as a paradigm shift not only in the plant sciences. Plant Sign Behav 2:205–207

Barlow PW (2008) Reflections on 'plant neurobiology'. Biosystems 92:132–147

Bell P (2001) Viral eukaryogenesis: was the ancestor of the nucleus a complex DNA virus? J Mol Evol 53:251–256

Bender J (2004) Chromatin-based silencing mechanisms. Curr Opin Plant Biol 7:521–526

Ben Jacob E (2004) Bacterial linguistic communication and social intelligence. Trends Microbiol l 2:366–372

Bonke M, Tähtiharju S, Helariutta Y (2005) Lessons from the root apex. In: Fleming AJ (ed) Intercellular Communication in Plants. Ann Plant Rev, Blackwell Publishing, Oxford, 16: 199–223

Braam J (2005) In touch: plant responses to mechanical stimuli. New Phytol 165:373–389

Brenner ED, Stahlberg R, Mancuso S et al. (2006) Plant neurobiology: an integrated view of plant signaling. Trends Plant Sci 11:413–419

Brenner ED, Stahlberg R, Mancuso S et al. (2007) Response to Alpi et al.: Plant neurobiology: the gain is more than the name. Trends in Plant Sci 12:285–286

Brundrett MC (2002) Coevolution of roots and mycorrhizas of land plants. New Phytol 154: 275–304

Buckley TN (2005) The control of stomata by water balance. New Phytol 168:275–292

Callaway RM (2002) The detection of neighbors by plants. Trends Ecol Evol 17:104–105

Campagnoni P, Blasius B, Nick P (2003) Auxin transport synchronizes the pattern of cell division in a tobacco cell line. Plant Physiol 133:1251–1260

Carol RJ, Dolan L (2006) The role of reactive oxygen species in cell growth: lessons from root hairs. J Exp Bot 57:1829–1834

Casson SA, Lindsey K (2003) Genes and signalling in root development. New Phytol 158:11–38

Coupland G (2005) Intercellular communication during floral initiation and development. In: Fleming AJ (ed) Intercellular Communication in Plants. Ann Plant Rev, Blackwell Publishing, Oxford, 16:178–197

Delledonne M (2005) NO news is good news for plants. Curr Opin Plant Biol 8:390–396

Denison FR, Kiers TE (2004) Why are most rhizobia beneficial to their plants, rather than parasitic? Microb Infect 6:1235–1239

Dessaux Y (2004). Biological Communications and Interactions in the Rhizosphere. Presented at "Rhizosphere 2004"-congress, 12–17. September 2004, Munich/Germany

De Vos M, Van Oosten VR, Jander G et al. (2007) Plants under attack. Plant Sign Behav 2:527–529

Dugas DV, Bartel B (2004) MicroRNA regulation of gene expression in plants. Curr Opin Plant Biol 7:512–520

Dunn AK, Handelsman J (2002) Toward an understanding of microbial communities through analysis of communication networks. Antonie van Leeuwenhoeck 81:565–574

Dunoyer P, Voinnet O (2005a) The complex interplay between plant viruses and host RNA-silencing pathways. Curr Opin Plant Biol 8:415–423

Dunoyer P, Voinnet O (2005b) RNA as a signalling molecule. In: Fleming AJ (ed) Intercellular Communication in Plants. Ann Plant Rev, Blackwell Publishing, Oxford, 16:49–83

Eigen M, Winkler R (1975) Das Spiel. Naturgesetze steuern den Zufall. Pieper, München

Engelberth J, Alborn HAT, Schmelz EY et al. (2004) Airborne signals prime plants against insect herbivore attack. Proc Natl Acad Sci U S A 101:1781–1785

Estabrock EM, Yoder JI (1998) Plant-plant communications: rhizosphere signalling between parasitic angiosperms and their hosts. Plant Physiol 116:1–7

Farmer E, Schulze-Lefert P (2005) Biotic interactions: from molecular networks to inter-organismal communities. Curr Opin Plant Biol 8:343–345
Federle MJ, Bassler BL (2003) Interspecies communication in bacteria. J Clin Invest 112: 1291–1299
Fleming AJ (2005) The plant extracellular matrix and signalling. In: Fleming AJ (ed) Intercellular Communication in Plants. Ann Plant Rev, Blackwell Publishing, Oxford, 16:85–107
Flores HE, Vivanco JM, Loyola-Vargas VM (1999) 'Radicale' biochemistry: the biology of root-specific metabolism. Trends Plant Sci 4:220–226
Friml J, Wisniewsky J (2005) Auxin as an intercellular signal. In: Fleming AJ (ed) Intercellular Communication in Plants. Ann Plant Rev, Blackwell Publishing, Oxford, 16:1–26
Gillespie T, Oparka KJ (2005) Plasmodesmata – gateways for intercellular communication in plants. In: Fleming AJ (ed) Intercellular Communication in Plants. Ann Plant Rev, Blackwell Publishing, Oxford, 16:109–146
Goh CH, Nam HG, Park YS (2003) Stress memory in plants: a negative regulation of stomatal response and transient induction of rd22 gene to light in abscisic acid-entrained Arabidopsis plants. Plant J 36:240–255
Golz JF (2005) Lessons from the vegetative shoot apex. In: Fleming AJ (ed) Intercellular Communication in Plants. Ann Plant Rev, Blackwell Publishing, Oxford, 16:147–177
Guerts R, Fedorova E, Bisseling T (2005) Nod factor signalling genes and their function in the early stages of Rhizobium infection. Curr Opin Plant Biol 8:346–352
Han S, Green L, Schnell DJ (2009) The signal peptide peptidase is required for pollen function in Arabidopsis. Plant Physiol 149:1289–1301
Hellmeier H, Erhard M, Schulze ED (1997) Biomass accumulation and water use under arid conditions. In: Bazzaz FA, Grace J (eds) Plant Resource Allocation. Academic Press, London
Hirsch AM, Bauer DW, Bird DM et al. (2003) Molecular signals and receptors: controlling rhizosphere interacting between plants and other organisms. Ecol 84:858–868
Hiscock SJ, Bright J, McInnis SM et al. (2007) Signaling on the stigma. Plant Sign Behav 2:23–24
Imaizumi-Anraku H, Takeda N, Carpentier M et al. (2005) Plastid proteins crucial for symbiotic fungal and bacterial entry into plant roots. Nature 433:527–531
Jablonka E, Lamb MJ (2002) The changing concept of epigenetics. Ann NYAcad Sci U S A 981:82–96
Jorgensen R (1993) The origin of land plants: a union of alga and fungus advanced by flavenoids. Biosystems 31:193–207
Jorgensen R (2004) Restructuring the genome in response to adaptive challenge: McClintock's bold conjecture revisited. Cold Spring Harbor Symposia on Quantitative Biology 69: 349–354
Kahmann R, Basse C (2001) Fungal gene expression during pathogenesis-related development and host plant colonization. Curr Opin Microbiol 4:374–380
Kant MR, Ament K, Sabelis MW et al. (2004) Differential timing of spider mite-induced direct and indirect defenses in tomato plants. Plant Physiol 135:483–495
Kempema LA, Cui Y, Holzer FM et al. (2007) Arabidopsis transcriptome changes in response to phloem-feeding silverleaf whitefly nymphs. Similarities and distinctions in responses to aphids. Plant Physiol 143:849–865
Keyes J, O'Malley R, Kim D et al. (2000) Signaling organogenesis in parasitic angiosperms: xenognosin generation, perception and response. J Plant Growth Regul 19:217–231
Kidner CA, Martienssen RA (2005) The developmental role of micro RNA in plants. Curr Opin Plant Biol 8:38–44
Kim I, Zambryski PC (2005) Cell to-cell communication via plasmodesmata during Arabidopsis embryogenesis. Curr Opin Plant Biol 8:593–599
Kovalchuk I, Kovalchuk O, Kalck V et al. (2003) Pathogen-induced systemic plant signal triggers DNA rearrangements. Nature 423:760–762
Kuhn T (1962) The Structure of Scientific Revolutions. The University of Chicago Press, Chicago
Kumar A, Bennetzen JL (1999) Plant retrotransposons. Annu Rev Genet 33:479–532

Kurata T, Okada K, Wada T (2005) Intercellular movement of transcription factors. Curr Opin Plant Biol 8:619–624
La Cerra P, Bingham R (2002) The Origin of Minds. Harmony Books, New York
Lammers PJ (2004) Symbiotic signalling: new functions for familiar proteins. New Phytol 16: 324–326
Lee A, Hirsch AM (2006) Signals and responses. Plant Sign Behav 1:161–168
Lolle SJ, Victor JL, Young JM et al. (2002) Genome-wide non-mendelian inheritance of extra-genomic information in Arabidopsis. Nature 434:505–509
Ma W, Smigel A, Tsai YC et al. (2008) Innate immunity signaling: cytosolic Ca^{2+} elevation is linked to downstream nitric oxide generation through the action of calmodulin or a calmodulin-like protein. Plant Physiol 148:818–828
Martin KC (2004) Local protein synthesis during axon guidance and synaptic plasticity. Curr Opin Neurobiol 14:305–310
Mauch-Mani B, Mauch F (2005) The role of abscisic acid in plant pathogen interactions. Curr Opin Plant Biol 8:409–423
McClintock B (1984) The significance of responses of the genome to challenge. Science 226: 792–801
McCubbin AG (2005) Lessons in signalling in plant self-incompatibility systems. In: Fleming AJ (ed) Intercellular Communication in Plants. Ann Plant Rev, Blackwell Publishing, Oxford, 16:240–275
Meagher TR, Vassiliadis C (2005) Phenotypic impacts of repetitive DNA in flowering plants. New Phytol 168:71–80
Mescher MC, Runyon JB, De Moraes CM (2006) Plant host finding by parasitic plants. Plant Sign Behav 1:284–286
Mithöfer A, Wanner G, Boland W (2005) Effects of feeding spodoptera littoralis on lima bean leaves. II. Continuous mechanical wounding resembling insect feeding is sufficient to elicit herbivory-related volatile emission. Plant Physiol 137:1160–1168
Moccia R, Chen D, Lyles V et al. (2003) An unbiased cDNA library prepared from isolated Aplysia sensory neuron processes is enriched for cytoskeletal and translational mRNAs. J Neurosci 23:9409–9417
Morita MT, Tasaka M (2004) Gravity sensing and signalling. Curr Opin Plant Biol 7:712–718
Müller CB, Kraus J (2005) Symbiosis between grasses and asexual fungal endophytes. Curr Opin Plant Biol 8:450–456
Nomura K, Melotto M, He SY (2005) Suppression of host defense in compatible plant-Pseudomonas syringae interactions. Curr Opin Plant Biol 8(4):361–368
Nürnberger T, Brunner F, Kemmerling B, Piater L (2004) Innate immunity in plants and animals: striking similarities and obvious differences. Immunol Rev 198:249–266
Pare PW, Tumlinson JH (1999) Plant volatiles as a defense against insect herbivores. Plant Physiol 121:325–331
Peak D, West JD, Messinger SM et al. (2004) Evidence for complex collective dynamics and emergent-distributed computation in plants. Proc Natl Acad Sci U S A 101:918–922
Pearce G, Ryan CA (2003) Systemic signalling in tomato plants for defense against herbivores: isolation and characterization of three novel defense-signalling glycopeptide hormones coded in a single precursor gene. J Biol Chem 278:30044–30050
Pearce G, Bhattacharya R, Chen YC (2008) Peptide signals for plant defense display a more universal ole. Plant Sign Behav 3:1091–1092
Pearson H (2005) Cress overturns textbook genetics. Nature 434:351–360
Press MC, Phoenix GK (2005) Impacts of parasitic plants on natural communities. New Phytol 166:737–751
Rapp RA, Wendel JF (2005) Epigenetics and plant evolution. New Phytol 168:81–91
Romero GQ, Benson WW (2005) Biotic interactions of mites, plants and leaf domatia. Curr Opin Plant Biol 8:436–440
Ross JJ, Wolbang CM (2008) Auxin, gibberellins and the gravitropic response of grass leaf sheath pulvini. Plant Sign Behav 3:74–75

Roossinck MJ (2005) Symbiosis versus competition in plant virus evolution. Nat Rev Microbiol 3:917–924
Ryan F (2002) Darwin's Blind Spot: Evolution Beyond Natural Selection. Houghton Mifflin Company, Boston
Samaj J, Baluska F, Voigt B et al. (2004) Endocytosis, Actin Cytoskeleton and Signalling. Plant Physiol 135:1150–1161
Samaj J, Read ND, Volkmann D et al. (2005) The endocytic network in plants. Trends Cell Biol 15:425–433
Schilmiller AL, Howe GA (2005) Systemic signalling in the wound response. Curr Opin Plant Biol 8:369–377
Schlicht M, Strnad M, Scanlon MJ et al. (2006) Auxin immunolocalization implicates vesicular neurotransmitter-like mode of polar auxin transport in root apices. Plant Sign Behav 1:122–133
Schultz JC, Appel HM (2004) Cross-kingdom cross-talk: hormones shared by plants and their insect herbivores. Ecol 85:70–77
Shackelton LA, Holmes EC (2004) The evolution of large DNA viruses: combining genomic information of viruses and their hosts. Trends Microbiol 12:458–465
Sharma A, Sahgal M, Johri BN (2003) Microbial communication in the rhizosphere: operation of quorum sensing. Curr Sci 85:1164–1172
Srinivas BP, Hülskamp M (2005) Lessons from leaf epidermal patterning in plants. In: Fleming AJ (ed.) Intercellular Communication in Plants. Ann Plant Rev, Blackwell Publishing, Oxford, 16:225–239
Strand A (2004) Plastid-to-nucleus signalling. Curr Opin Plant Biol 7:621–625
Tassetto M, Maizel A, Osorio J et al. (2005) Plant and animal homeodomains use convergent mechanisms for intercellular transfer. EMBO Rep 6:885–890
Teplitski M, Robinson JB, Bauer WD (2000) Plants secrete substances that mimic bacterial N acylhomoserine lactone signal activities and affect population density dependent behaviors in associated bacteria. Mol Plant Microbe Interact 13:637–648
Thompson KR, Olofsdotter OK, Chen DY et al. (2004) Signaling from synapse to nucleus during long-term synaptic plasticity: a role for the classical active nuclear import pathway. Neuron 44:997–1009
Thummel CS, Chory J (2002) Steroid signaling in plants and insects – common themes different pathways. Genes Dev 16:3113–3129
Tomilov AA, Tomilova NB, Abdallah I et al. (2005) Localized hormone fluxes and early haustorium development in the hemiparasitic plant Triphysaria versicolor. Plant Physiol 138:1469–1480
Torres MA, Dangl JL (2005) Functions of respiratory burst oxidase in biotic interactions, abiotic stress and development. Curr Opin Plant Biol 8:397–403
Trewavas A (2003a) Aspects of plant intelligence. Ann Bot 92:1–20
Trewavas A (2003b) Aspects of plant intelligence: an answer to firn. Ann Bot 93:353–357
Trewavas A (2005) Green plants as intelligent organisms. Trends Plant Sci 10:413–419
Trewavas A (2007) Response to Alpi et al.: Plant neurobiology – all metaphors have value. Trends Plant Sci 12:231–233
Tripathi SK, Tuteja N (2007) Integrated signaling in flower senescence. Plant Sign Behav 2:437–445
Turkington R, Sackville Hamilton R et al. (1991) Within-population variation in localized and integrated responses of Trifolium repens to biotically patchy environments. Oecologia 86:183–192
Vandenkoornhuyse P, Baldauf S, Leyval C et al. (2002) Extensive fungal diversity in plant roots. Science 295:2051
Van der Putten WH, Vet L, Harvey JA et al. (2001) Linking above- and belowground multitrophic interactions of plants, herbivores, pathogens, and their antagonists. Trends Ecol Evol 16:547–554
Van West P, Morris BM, Reid B et al. (2002) Oomycete plant pathogens use electric fields to target roots. Mol Plant Microbe Interact 15:790–798

Villarreal LP (2005). Viruses and the Evolution of Life. American Society for Microbiology Press, Washington

Walker TS, Bais HP, Grotewold E et al. (2003) Root exudation and rhizosphere biology. Plant Physiol 132:44–51

Wang MB, Metzlaff M (2005) RNA silencing and antiviral defense in plants. Curr Opin Plant Biol 8:216–222

Weigl D, Jürgens G (2005) Genetics. Hotheaded healer. Nature 434:443

Wen F, VanEtten HD, Tsaprailis G et al. (2007) Extracellular proteins in pea root tip and border cell exudates. Plant Physiol 143:773–783

Witzany G (1995) From the "logic of the molecular syntax" to molecular pragmatism. Ev Cog 1:148–168

Witzany G (2005) Natural history of life: history of communication logics and dynamics. SEED J 5:27–55

Witzany G (2006) The *Logos* of the *Bios* 1. Contributions to the Foundation of a three-leveled Biosemiotics. Umweb, Helsinki

Xia Y (2005) Peptides as signals. In: Fleming AJ (ed) Intercellular Communication in Plants. Ann Plant Rev, Blackwell Publishing, Oxford, 16:27–47

Xonocostele-Cazares B, Xiang Y, Ruiz-Medrano R et al. (1999) Plant paralog to viral movement protein that potentiates transport of mRNA into the phloem. Science 283:94–98

Yoder JI (1999) Parasitic plant responses to host plant signals: a model for subterranean plant-plant interactions. Curr Opin Plant Biol 2:65–70

Zhang S, Cai Z, Wang X (2009) The primary signalling outputs of brassinosteroids are regulated by abscisic acid signaling. Proc Natl Acad Sci U S A doi: 10.1073/pnas.0900349106

Zimmermann U, Schneider H, Wegner LH et al. (2004) Water ascent in tall trees: does evolution of land plants rely on a highly metastable state? New Phytol 162:575–615

Zipfel C, Felix G (2005) Plants and animals: a different taste for microbes? Curr Opin Plant Biol 8:353–360

Zyalalov AA (2004) Water flows in higher plants: physiology, evolution and system analysis. Russ J Plant Physiol 51:547–555

Chapter 3
Communicative Competences of Honey-Bees

Abstract Like ants, bees derive evolutionarily from parasitoid wasps, the most numerous species in the animal kingdom with nearly 100,000 different species. Like other flying insects, bees are important pollinators for plants, which would not be able to produce fruits without them. Since Karl von Frisch's work it has been evident that the highly complex social behaviour of bee swarms is organised and coordinated by sign-mediated interactions, i.e. communication. If communication processes are disturbed this may have fatal consequences for bee colonies. As in every other natural language the same sign sequences may have different meanings in different contexts. This means that bees with a limited repertoire of signs can transport different messages via identical semioses which trigger different response behaviours with far-reaching consequences. As in every other natural language, bee languages also differ in habitat-dependent dialects. The language of honey bees in colder hemispheres is the only known non-human language which uses body movements that represent symbolic meaning functions.

3.1 Introduction

Communication serves to coordinate action and behaviour as well as to form associations between linguistic-competent individuals. The use of linguistic signs takes place in specific language media, whereby actions or behaviours (in sequences of actions or behaviours) can also assume sign-like character. This sign-use takes place in regulative, constative and – less frequently – in generative linguistic action as well, or in linguistic behaviour that is also characterised by regulative, constative and generative features. Without sign-use there would be no coordination processes.

The rules that underlie or govern linguistic behaviour or linguistic action stem largely from the practice of *social interactions*. Nevertheless there are some indicators that interpretation processes occur not only according to swarm vectors but also to individual needs (Grüter et al. 2008).

The individual's genetic make-up gives it the ability to communicate in a species-specific environment. Additionally it is necessary to learn and remember real-life

G. Witzany, *Biocommunication and Natural Genome Editing*,
DOI 10.1007/978-90-481-3319-2_3, © Springer Science+Business Media B.V. 2010

experiences which orientate inherited abilities by concrete social interactions, i.e. swarm behavioural patterns.

The inter-organismic communication we will discuss here demonstrates the rule-governed, sign-mediated interaction between conspecifics. The increased skill in following these rules goes hand in hand with the improved ability to use linguistic signs within rule-governed interactions between individuals of a real, species-specific, life-world (*Lebenswelt*). Living 'Beeings' in general are unaware of these underlying rules nor are they able to explicate the rules as rules. Only humans are able to identify and explicitly reflect on these levels of rules.

The language and communication of the honey-bee, which has been studied in great detail, can serve as an excellent example of non-human language (Frisch 1952, 1953, 1955, 1965, 1970, 1971; Lindauer 1975, 1981; Seeley 1982, 1992, 1995; Heinrich 1981). This can be illustrated by two cases in which communication, coordination of behaviour, and the formation of associations are achieved through linguistic signs.

In contrast with the investigations on language and communication in the other organismic kingdoms outlined in this book, I will focus in the case of the honey-bees only at the inter-organismic level, i.e. sign-mediated interactions between the same or related species. This is because, in contrast with all other cases of biocommunication we can mention, here we have the rare phenomenon that body moving patterns act as symbols (Frisch 1953; Sherman and Visscher 2002). In contrast to indexical or iconic sign-use symbols do not represent by themselves what they mean, but are a kind of natural convention. Their correct use must be trained within in vivo social interactions.

3.2 Honey-Bees in the Colder Hemispheres

Honey-bees originally stem from the warm regions of the Earth. The extension of their range into the northern hemisphere brought with it the problem of how to deal with longer cold periods. Winter requires the development of a specific survival strategy that was unnecessary in the geographic and phylogenetic origins of honey-bees.

Searching for and finding suitable over-wintering sites are critical for the survival of honey-bees in temperate and more northern latitudes. The complex communication and behavioural coordination of individuals in this community requires correspondingly differentiated communication abilities and skills; without these, no suitable housing could be selected. A mistake in the selection of a hive leaves no opportunity for a second attempt: the correct choice is a life-or-death situation for the bee colony. Today we know that only one-quarter of all newly-established bee colonies survive the first winter (Seeley 1982). Once the bees survive the first winter in a well-chosen site, the probability of surviving for another five years is high. How does the selection of an appropriate site take place?

3.2.1 *The Communication Process Behind the Founding of a New Colony*

In the cold season, the bee colony forms a tight aggregation in its hive. A great number of bees join to form a type of outer shield; fine quivering movements of their flight muscles help maintain an ambient temperature of at least 10°C in the colony. Abundant supplies of honey ward off starvation. Spring marks the beginning of an intensive phase of brood rearing, and newly-emerging bees lead to dense colonies and reduced space in the hive. Precisely this condition is a sign for the worker honey-bees to construct queen cells in which a number of future queens can be reared simultaneously (Seeley 1982). The old queen uses sound signs to communicate with the enclosed future queens. If one imitates the sounds of the enclosed future queens with the appropriate instruments, then the old queen answers these artificially produced sounds quite specifically. Although the queen is known to communicate with the future queens, the subject of the communication remains unknown. Before the replacements emerge and decide the future leadership of the colony in a stinging duel, the mother queen leaves the nest with half the original colony.

The initial flight is rather short, in any case less than 50 m (Seeley 1982). The queen alights on a more or less suitable object and is immediately surrounded by a cluster of bees. As soon as the cluster is completed, a few hundred so-called scouts swarm out to search the terrain for a suitable new hive. These scouts are the oldest bees in the colony, i.e. those that have already collected food for the original colony and are therefore already familiar with the surroundings (Thom 2003; Menzel et al. 2006).

The selection criteria for the new home are quite differentiated (Seeley 1992; Lindauer 1975). The height of the entrance hole must lie at least 2 m above the ground in order effectively to rule out any disturbance of the colony by other animals. The opening of the hole should be no larger than 50 cm in order to permit reliable regulation of the hive's internal climate in winter, even if the temperatures drop to very low values. The hole should also face south: this enables the bees to swarm out and defecate at even the slightest outside temperature increases in winter. The volume of the entire hive should not be fewer than 100 l in order to provide sufficient room for the honey stores the colony needs to over-winter; at the same time, volumes in excess of 100 l make the regulation of the inside temperature difficult. The presence of old honeycombs is a positive criterion because it considerably reduces or even eliminates the time- and energy-consuming effort of honeycomb construction (Seeley 1992; Lindauer 1975).

Each of the scouts that have swarmed out returns to the colony as soon as she has found a site that seems suitable. She lands in the swarm and begins to carry out the characteristic movements that researchers have termed dances (Lindauer 1975). Such dances are displayed not only during the search for new hives, but also in locating suitable feeding sites. The dances have a communicatory character and represent linguistic signs whose expression enables comprehensible information to be relayed to other bees. Because the scouts do not return with pollen or nectar, the dance is not a message about feeding sites but rather about where and in which direction a suitable place to build a hive can be found. These so-called waggle dances resemble a figure-of-eight in whose central section the abdomen is waggled. The waggle dance can be defined as constative linguistic behaviour (Lindauer 1975). The direction of the central section of the danced figure-eight points to the direction of the new site in relation to the respective position of the sun (Lindauer 1975; Frisch 1965).

The greater the distance to the prospective site, the longer the waggling motion for that particular stretch lasts. This is interconnected with increasing inexact message transfer (Beekman et al. 2005). Three or four other scouts observe the waggle dance from close quarters. Each scout advertises her discovery with her own dance. The more suitable she believes her discovery to be, the more vigorous the dance. The greater a scout's doubt about her own discovery, the more subdued her dance. The latter are quickly attracted to the dances of their more agitated neighbours and follow up on the communication of one such dancing bee: they fly to the site indicated by the dance and inspect the keenly promoted hive. After the inspection, these scouts return to the swarm: each scout that is more convinced of the new hive than of the one she originally found begins an agitated dance to promote the new site. One after the other, all sites that were keenly promoted by scouts are visited by the other scouts. This consensus-building process gradually leads to agreement on one site (Frisch 1965; Lindauer 1975; Seeley 1982; Visscher 2007).

It should be mentioned that the inspection of a potential hive site is quite a precise process: the bee walks up and down the entire hollow, often covering a distance of 50 m (Seeley 1982). This allows her to cover the entire inner surface of the cavity. When a scout gives up her original, first discovery and *consensually* agrees with the other scouts on another site, then the experienced scouts are in true agreement (Seeley 1982).

The new beehive can be up to 10 km away from the original site. The search lasts no more than three to four days. If no suitable place is found, then the bees begin constructing a hive directly at the first landing site and, since such hives cannot withstand the winter weather, the colony dies during the first cold spell.

If, on the other hand, the scouts have agreed on a new site, they force their way to the surface of the bee cluster in a zigzag course. At this point the entire colony begins to beat its wings in order to raise its temperature to 36°C (Seeley 1982; Heinrich 1981). This is a necessary precondition for the bees to be able to fly after this quiescent state (Seeley et al. 2003). If the colony is knocked off-balance before this temperature is reached, the whole swarm falls to the ground. When, after a few minutes, the correct temperature is reached, the scouts give the sign to take off: they force the aggregated bees apart in a series of so-called buzzing runs. The entire swarm disperses and ascends into the air, forming a cloud of bees with a diameter of approximately 10 m. Within this cloud, the scouts repeatedly take off in the direction of the new site, thereby showing the other bees the correct bearing of the destination. In the first 30 m the swarm makes only very slow progress, but it picks up speed dramatically over the next 200 m.

Upon reaching the newly-selected site, the scouts emit a sign (Frisch 1965). The nature of this sign is still unknown. The swarm reacts to this sign, however, and comes to a standstill above the new hive. The scouts drop from the stationary swarm, alight on the entrance of the new home, and mark it precisely with a scent. Shortly thereafter the entire swarm takes over the hive. Within hours they remove all dirt, begin immediately to build the combs, and fly in search of pollen and nectar. This marks the end of this specific communication process until the following year.

3.2.2 The Sign-Mediated Interaction of Foraging

The second display of honey-bee language that I describe here picks up where the first left off. It plays an equally important role in enabling the bee colony to survive over the winter.

So-called foraging bees are responsible for finding suitable food sources. Foraging bees are experienced and possess memory (Menzel et al. 2006; Towne 2008). Upon finding such a site, the forager returns to the surface of the hive and begins her own waggle dance (Rohrseitz and Tautz 1999). She brings along pollen and nectar, which not only informs the other bees that the dance refers to a feeding site and not to a new hive site (Frisch 1955, 1970), but also provides information on the quality and quantity of the food (Farina et al. 2005). The linguistic signs and the sequence of these signs are the same as in the first communication process described above (in which only scouts were involved). In this case, however, the waggle dance sequences are relevant to all encountered foragers, prompting them to collect food in the described direction and described distance. The constative linguistic behaviour has changed into regulative linguistic behaviour. The linguistic behaviour is different, the communication processes pursue different goals, yet the linguistic signs that are employed have remained the same. Additionally the food-gathering honey-bees now begin to produce special volatiles to recruit other honey-bees (Tautz 1996; Tautz and Rohrseitz 1998; Dyer 2002; Thom 2003; De Marco 2005; Thom et al. 2007).

The above scenario refers only to information on feeding sites that are more than 25 m away. The bees again dance a figure-of-eight. The orientation of the central section of the figure signals the direction of the feeding site in relation to the position of the sun. If the dance takes place on a vertical honeycomb, the deviation in direction between the feeding site and the sun's position is accurately recreated as the deviation from the vertical (Frisch 1965). As in the case of the first communication process, distances are depicted temporally: the waggling in the central section of the figure lasts longer for longer stretches than it does for short ones. Thus, waggling for 1 s can indicate a distance of 500 m, while waggling for 2 s can indicate 2 km. Other bees follow the waggle-dancing bee at close quarters, and certain odours provide additional information about the site. Rather than approaching such places directly, the individual bees take small detours. They orientate themselves according to distinctive landscape features (Srinivasan et al. 1996). These orientations are determined in an arbitrary manner and are specific to the individual bee: they are not communicated to the others. If certain orientational features are experimentally altered, some bees can briefly become disoriented.

Although honey-bees in most cases are diurnal – most of them are colour-blind in moonlight – one species which flies on moonless nights has been observed. It recognises landmarks in starlight (Somanathan et al. 2008).

3.3 Further Features of Honey-Bee Communication

3.3.1 The Types of Dances and Their Meanings

To date, nine different dance types have been identified as linguistic signs (Frisch 1965):

1. The round-dance is a call to search for food in all directions within a radius of 25 m.
2. The waggle-dance describes the direction of the destination in terms of the respective position of the sun and defines the distance.
3. The tremble-dance describes a conspicuous type of movement made by successfully returning foragers. They hastily make their way across the honeycomb, bumping into colony members and informing them that something is going on, e.g. that food is available.
4. The ruck-dance is carried out by foragers that are emptying their honey sacs and involves intermittent, directed tail wagging. It serves more to indicate a general dancing mood than to impart any specific message.
5. The sickle-dance has been observed in every bee species (with one exception) in the transition between the round-dance and the waggle dance (figure-of-eight). The opening of the 'sickle' in the dance pattern denotes the direction to the feeding site.
6. The buzzing run is the sign to disperse. Scouts barge through the interlocked bees in the swarm in an undirected, zigzag course and audibly buzz their wings.
7. In the 'cleaning run' the bee shakes its body from one side to the other.
8. In the vibration-dance, one bee takes up contact with another, whereby it rapidly vibrates its abdomen. The meaning of this dance has not yet been deciphered, although there is strong evidence that it involves a communication form combining dance and acoustic signals.
9. Finally, the jitter-dance is an expression of neurotic behaviour and is disregarded by the surrounding bees. Research has shown it to be a result of a traumatic experience such as severe impact, poisoning, injury to appendages, or extreme state of alarm.

The type of sign use that we designate as dances is a genetically acquired linguistic competence: even without prior socialisation, i.e. the presence of older bees, juveniles develop the ability to dance. These kinds of dances are not able to transport messages. Subsequent social interaction with bees of the same age is important to develop meaningful dances: carrying out linguistic behaviour and heeding the calls for specific action require some degree of practice and experience in participating in mutual interactions.

While the ability of bees to take their bearings according to the respective position of the sun is innate, the specific skills are gained and perfected in the course of a few days of flying experience. Interestingly, bees recognise the sun as having a 24-hour course, so that they can carry out their dance at the correct angle vis-à-vis the sun even in the dark (Frisch 1965; Lindauer 1975).

3.3.2 Forms of Communication Beyond Dances

Honey-bees exhibit other forms of communication that are either combined with or separate from the communicatory dances themselves. Various wing-beating frequencies or abdominal vibrations can transmit movement frequencies on suitable substrates (Kirchner 1993); other bees are able to identify their meaning. This is evident in the specific reactions to certain frequencies (Frisch 1965).

Odours, which are actively employed, are apparently an even more significant form of danceless communication (Frisch 1965, 1970). During her nuptial flight, the queen bee emits scents that attract the males (drones). The bees positioned at the entrance hole of the hive beat their wings to waft the smell of their scent glands toward the arriving workers and thus guide them to the entrance. The scent glands are also used to mark certain food sources, enabling other bees to find these sites more easily (Breed 1998, Schmitt et al. 2007). Bees that are threatened or attacked extend their sting and whirr their wings to exude an alarm scent which is not identical with bee poison. The alarm induces members of the colony to attack. This attack is generally directed at moving objects in the vicinity.

3.3.3 Humans can Understand the Bee-Language

Once one understands the bees' communicative behaviour and their use of linguistic signs at the syntactic, semantic, and pragmatic level, it should be possible to identify the practical meaning of the information content. In fact, bee researchers, by observing the dances, have been able to locate the feeding sites down to the metre! The only deviations pertained to direction, not to distance; humans are less adept than the studied bees at identifying the bearing in relation to the sun's position and potential crosswinds (Frisch 1965).

3.3.4 Dialects of the Bee-Language

Bee colonies form relative language communities that are distinguished by dialects (Sen Sarma et al. 2004). Experimentally-mixed colonies of Austrian and Italian bees revealed clear differences in the interpretation of the dance tempo, which indicates the distance to the feeding site. When the Austrian bees communicated a suitable feeding site at a distance of 300 m, for example, the Italian bees executed the instruction in exactly the right direction, yet over a distance of 500 m. Vice versa, a 200-m dance by the Italian bees meant a much shorter distance to the Austrian bees. Thus, despite identical rules being applied to the same linguistic signs, distinct differences existed in the meaning of the signs (Frisch 1965). Interestingly, these differences in bee language dialects are even compatible over longer time distances. It depends on the capability of social learning of the bee populations. Longer time enables processes of training of different meanings of identical moving patterns (Su et al. 2008).

In stingless bees, the use of symbolic signs in regulative and constative linguistic behaviour to indicate direction and distance is not developed. These bees must accompany and guide inexperienced conspecifics to every discovered food site.

3.4 Language and Communication in Bees: Context Determines Meaning

Twenty years before Karl von Frisch received a Nobel prize for his research into bee language, he was embroiled in a controversy involving so-called animal languages. He was accused of improperly using the term language to describe specific behavioural feats. His opponents argued that a very simple form of animal communication was involved, but certainly not language. In his response (Frisch 1953), Frisch proves that it is justifiable to speak of the language of bees because a system of signs is involved.

In this chapter I have discussed only two of many sign-mediated communication processes that can serve as examples of rule-governed, sign-mediated interactions between individual bees in a colony. Each of these communication processes encompasses a series of characteristic sign uses and sign combinations, whereby the context of usage clearly determines the meaning of the utilised sign sequences. Furthermore, various forms of behaviour evidently take on sign character and, when combined, can take on meaning and be understood as signs. The habitat specificity with which such language communities apply their stock of signs is reflected in the different dialects of bee communities.

The communication of honey-bees (a) with one another and (b) about something is necessary to exchange information, coordinate behaviour, and form associations between individuals of such social animal communities (Lindauer 1975). Survival without the sign-mediated interactions described above would be impossible. At the same time, this example vividly illustrates how certain behaviour can take on sign character within behaviour sequences.

Beyond using linguistic signs in regulative and constative linguistic behaviour, bees must originally have been able to perform generative linguistic behaviour, thereby ultimately constituting new life-forms. In order to survive the winter, the swarm first had to have selected the correct over-wintering site through appropriate communication. At some point, the rules underlying this communication were innovatively generated, much in the same way as the transition from signs with mere reference character (honey-bees of warmer latitudes) to symbolic sign systems (northern hemisphere bees) must have been a marked innovative step. The ability to survive the winter eventually became fixed in the genetic text of these surviving bee generations. We know that this happened, but can only guess at how it happened.

The survival strategy of honey-bees in colder latitudes clearly shows that communication experience in generative linguistic behaviour, which substantially and permanently supersedes the originally innate language competence, can constitute an expanded communication competence. This competence is hereditary in an expanded (or at least modified) form; in the context of social animals, it can be differentiated as broadened language play skills, for example, through learning processes (Lindauer 1975). Learning processes depend on the capability to memorise and short-term as well as long-term memory has been proven (Menzel and Muller 1996).

3.4.1 Foundation of a New Colony

As demonstrated in our treatment of two sign-mediated communication processes in the language of northern hemisphere honey-bees, in certain situations the behavioural context determines the meaning of the linguistic signs used. The bees' ability to interact socially is no doubt genetically fixed. The constitution of the specific performance, however, i.e. of the actual communication process, is contingent on the actual situational demand.

In the sign-mediated communication process underlying the foundation of a new colony, only scouts participate in the search for a new home. They are the oldest bees in the swarm and have already gathered food for the parent hive; they are fully familiar with the features of the local terrain. Why do only these experienced scouts swarm out and not the inexperienced ones as well? Does the flight of the queen cause certain genetic text sequences in the scouts to be expressed, i.e. those that code for and initiate such behaviour? Or does the rule governing the participation of experienced scouts alone underlie some other species-specific, intersubjective communication?

The criteria that a prospective hive must fulfil are so differentiated that one can reasonably assume a genetically determined inspection and evaluation behaviour. On the other hand, these evaluation criteria clearly do not exist from the onset: they must have been constituted by experience, followed by subsequent genetic fixation. Pragmatic situations formed the evaluation pattern for epigenetic imprinting which seemed to be genetically encoded later on. Naturally, there is no reason to doubt that natural genome editing agents competent in genetic text processing and integration carried out this fixation, i.e. have structured and, above all, inserted the respective sequence at the appropriate site in the genome.

No haphazard change or deformation of genetic text sequences can shape the highly differentiated selection criteria for the winter hives of northern hemisphere honey-bees: they are simply too rigorous. The failure of the hive selection process to match the required hive features closely can kill off the entire swarm in one winter. The argument that this involves the natural selection of many chance mutations would imply the extinction of all northern hemisphere bee populations before they ever had the opportunity to develop sufficiently differentiated selection criteria for suitable winter hives.

As demonstrated earlier, the process by which a potential winter home is scrutinised is itself incredibly complex and exact. The bees pace the entire length and breadth of the new site: not a millimetre is left out. This explains why a single bee covers a distance of nearly 50 m in the course of this inspection, even though the cavity itself is relatively small.

The sign-mediated communication process underlying the founding of a new bee colony also points to numerous other pragmatic situations that must be or, if they are epigenetically imprinted and even genetically fixed, must have been vital for the evaluatory function. The consultation between scouts about the potentially most suitable new home – in this case the tail waggle dance – raises the question: what induces bees that have identified a potential site as being less satisfactory to dance less vigorously, and bees that have identified a site as being highly suitable to dance more vigorously and to 'symbolically code' (Todt 1986) the direction and distance of their discovery? What induces the less lively dancers, those who are less convinced of their discovery, to take up the invitation of the more vigorously dancing bees to inspect the site they consider to be particularly suitable, especially when this involves repeating the same complex and time-consuming inspection procedure? What subsequently enables these bees to decide in favour of the recommended, inspected, and perhaps more highly-evaluated site and themselves promote this site with an appropriately intense dance (Seeley and Visscher 2008)? Furthermore, this new decision may itself be temporary, and another, even better, home may trigger a renewed inspection process, etc. At any rate, the final decision is a consensual decision by all scouts, all of whom have by then inspected the most highly-advocated home. If no consensus can be reached, no decision is taken and the swarm freezes to death at the site of their deliberations during the first cold spell.

Provided that the decision-making process represents sign-mediated communication, then it cannot be of the algorithmic type; rather, it must be a truly communicative process between conspecifics in a shared life-world (*Lebenswelt*). They represent subjects for one another because they use the same linguistic signs in the same sign-mediated communication process to achieve understanding, form associations, and coordinate behaviour. The fact that language is involved, i.e. language and not merely a formal procedure, opens the potential for generative and therefore entirely new linguistic behaviour. Otherwise, colder hemisphere bees would never have been able to differentiate the necessary sign-mediated communication processes (processes outside the repertoire of warmer hemisphere bees). Whereas southern hemisphere bees use behaviour to constitute signs with direct indicatory or invitational character, northern hemisphere bees employ movements to constitute and utilise a symbolic sign character for these movements; understanding these signs permits more differentiated messages to be deciphered (messages that even humans can understand, provided that they can determine the rules underlying the use of these movement signs).

Todt, a sociobiologist whose research was instrumental in initiating an interdisciplinary dialogue with semiotics in Germany, expressly underlines the use of symbols by bees of the colder hemisphere.

The specific sign-mediated communication process involved in searching for a home is terminated only when consensus has been reached. The process is completed when a new home (one selected exclusively by scouts) is inhabited and developed.

3.4.2 Food Gathering

This marks the onset of the second sign-mediated communication process described above – food-gathering. Again, the waggle dance is used to convey information. The rules underlying the movement sequences as well as the indication of direction and distance remain the same as in the preceding example. The sequence of signs is also the same. Their meaning, however, is different because they take on new meaning within the pragmatic context of a new communication process. The waggle dance may well be a rule-governed, genetically fixed behaviour that is expressed as the need arises: nonetheless, the actual situation in which the signs are used within a population of communicating conspecifics lends meaning to the signs themselves and determines their sequence in a dance.

In addition, the target group addressed by these expressions is not the same as in the preceding case. All foragers, not just the scouts alone, are called upon to search for food sites. One situation-specific feature is responsible for the fact that foragers (and not just scouts) are being addressed, even though the mode of expression and the utilised linguistic signs are the same as in the previous example in which scouts were prompted to swarm out: only when the dancers carry flower pollen – which is not the case when the task involves searching for a new hive – is the call valid for foragers as well. In the absence of pollen, the foragers do not react to the messages or invitations. Understanding (*Verständigung*) between bees is not limited to dance movements alone. These movements are combined with (the very important) vibratory movements (Kirchner and Towne 1994) of the wings and abdomen along with the rule-governed use of olfactory signs. This marks the limits of our comprehension of the bee language. Human beings can never hope to progress much beyond a passable understanding of the rules governing the bees' use of linguistic signs: beyond a certain complexity of sign combinations, mastering the specific modes of use would require becoming involved in the bees' communication process as interactional subjects. This inherently transcends human capabilities and points to the limits in the compatibility of trans-specific forms of communication, for example, in trans-organismic communication.

3.4.3 Dialects in Different Cultural Life-Worlds

One final pragmatic criterion for the signifying function of the utilised linguistic signs deserves mention: the occurrence of various bee dialects. The same sign (or the same sign sequence) can exhibit slightly different rules of usage in bee colonies that are geographically widely separated yet belong to the same species. In the case of the Austrian and Italian bees described earlier, the form in which the same symbolic (behavioural) sign is expressed can translate into site deviations of several hundred metres. The pragmatic context, in this case the bee colony's actual life-world (*Lebenswelt*), determines the semantic rules according to which this sign is interpreted. As we know now, these rules which depend on cultural customs of honey-bee populations are subject to learning and memory capabilities. After a certain time-span honey-bees can change their sign using practice according to new situational contexts (Su et al. 2008).

References

Beekman M, Doyen L, Oldroyed BP (2005) Increase in dance imprecision with decreasing foraging distance in the honey bee Apis mellifera L. is partly explained by physical constraints. J Comp Physiol A 191:1107–1113

Breed MD (1998) Recognition pheromones of the honey bee, Bioscience 6:463–470

De Marco RJ (2005) How bees tune their dancing according to their colony's nectar influx: re-examining the role of the food-receivers' 'eagerness'. J Exp Biol 209:421–432

Dyer FC (2002) The biology of the dance language. Ann Rev Entomol 47:917–949

Farina WM, Grüter C, Diaz PC (2005) Social learning of floral odours inside the honeybee hive. Proc R Soc B 272:1923–1928.

Frisch KV (1952) Die wechselseitigen Beziehungen und die Harmonie im Bienenstaat. In: Colloques d. Centre Nat. de la Recherche Scient. XXXIV, Paris.

Frisch KV (1953) Sprache' oder, Kommunikation' der Bienen? In: Psychologische Rundschau 4, Amsterdam.

Frisch KV (1955) Sprechende Tänze im Bienenvolk. Verlag der Bayerischen Akademie der Wissenschaften, München.

Frisch KV (1965) Tanzsprache und Orientierung der Bienen. Springer, Berlin/Heidelberg/ New York.

Frisch KV (1970) Ausgewählte Vorträge 1911–1969. BLV, München/Basel/Wien.

Frisch KV (1971) Bees: Their Vision, Chemical Senses and Language, Revised ed. Cornell University Press, Ithaca.

Grüter C, Balbuena MS, Farina WM (2008) Informational conflicts created by the waggle dance. Proc Biol Sci 275:1321–1327

Heinrich B (1981) The mechanisms and energetics of honeybee swarm temperature regulation. J Exp Biol 91:25–55

Kirchner WH (1993) Vibrational signals in the tremble dance of the honeybee, Apis mellifera. Behav Ecol Sociobiol 33:169–172

Kirchner WH, Towne WF (1994) Die Tanzsprache der Bienen: eine akustische Kommunikation. Spektrum der Wissenschaft 8:68–75

Lindauer M (1975) Verständigung im Bienenstaat. Gustav Fischer Verlag, Stuttgart

Lindauer M (1981) Kommunikation im Bienenstaat - Arbeitsteilung, Nahrungssuche, Wohnungssuche. In: Scharf JH, Kämmerer W (eds) Nova Acta Leopoldina, Leipzig.

Menzel R, De Marco RJ, Greggers U (2006) Spatial memory, navigation and dance behaviour in Apis mellifera. J Comp Physiol A 192:889–903
Menzel R, Muller U (1996) Learning an memory in honeybees: from behaviour to neural substrates. Ann Rev Neurosci 19:379–404
Rohrseitz K, Tautz J (1999) Honey bee dance communication: waggle run direction coded in antennal contacts? J Comp Physiol A 184:463–470
Schmitt T, Herzner G, Weckerle B et al. (2007) Volatiles of foraging honeybees Apis mellifera (Hymenoptera: Apidae) and their potential role as semiochemicals. Apidologie 38:164–170
Seeley TD (1982) How honeybees find a home. Sci Am, 10, 144 ff.
Seeley TD (1992) The tremble dance of the honey bee: message and meanings. Behav Ecol Sociobiol 31:375–383
Seeley TD (1995) The Wisdom of the Hive. Harvard University Press, Cambridge (Massachusetts)
Seeley TD, Kleinhenz M, Bujok B et al. (2003) Thorough warm-up before take-off in honey bee swarms. Naturwissenschaften 90:256–260
Seeley TD, Visscher PK (2008) Sensory coding of nest-site value in honeybee swarms. J Exp Biol 211:3691–3697
Sen Sarma M, Esch H, Tautz J (2004) A comparism of the dance language in Apis mellifera carnica and Apis florae reveals striking similarities. J Comp Physiol A 190:49–53
Sherman G, Visscher PK (2002) Honeybee colonies achieve fitness through dancing. Nature 419:920–922
Somanathan H, Borges RM, Warrant EJ et al. (2008) Nocturnal bees learn landmark colours in starlight. Curr Biol 18:996–997
Srinivasan MV, Zhang SW, Lehrer M et al. (1996) Honeybee navigation en route to the goal: visual flight control and odometry. J Exp Biol 199:237–244
Tautz J (1996) Honey bee waggle dance: recruitment success depends on the dance floor. J Exp Biol 199:237–244
Tautz J, Rohrseitz K (1998) What attracts honey bees to a waggle dancer? J Comp Physiol A 183:661–667
Thom C (2003) The tremble dance of honey bees can be caused by hive-external foraging experience. J Exp Biol 206:2111–2116
Thom C, Gilley DC, Hooper J et al. (2007) The scent of the waggle dance. PloS Biol 5:1862–1867
Todt D (1986) Hinweis-Charakter und Mittler-Funktion von Verhalten. Zeitschrift für Semiotik 8:183–232
Towne WF (2008) Honeybees can learn the relationship between the solar ephemeris and a newly-experienced landscape. J Exp Biol 211:3737–3743
Su S, Cai F, Si A et al. (2008) East learns from west: asiatic honeybees can understand dance language of european honeybees. PloS ONE 3(6): e2365. doi:10.1371/journal.pone.0002365
Visscher PK (2007) Group decision making in nest-site selection among social insects. Ann Rev Entomol 52:255–275

Chapter 4
Biocommunication of Corals

Abstract Scleractinian corals are keystone species of tropical reef ecosystems. Development and growth of these organisms rely on complex and efficient communication patterns. Such patterns are sign-mediated interactions that cannot be reduced to mere exchange of information, but rather require active coordination and organisation. Communication processes within and among corals takes place at varying levels and occurs (a) within cells, (b) among cells, (c) between corals of the same species (conspecies), (d) between corals of different species, and (e) among trans-specific organisms that include all phyla of all kingdoms of shallow-water tropical ecosystems.

4.1 Introduction

Hitherto most specialists have considered corals as archaic, sessile animals that are placed at the lower level of organismic complexity. Now, however, it is becoming increasingly obvious that their evolution, growth and development depend upon complex and successful communication processes. This article gives an overview of the manifold levels of coral communication and thereby broadens the understanding of these organisms.

Even though tropical coral reefs are embedded in an oligotrophic setting, they are still capable of sustaining a lush environment. The frequent description of reefs as the 'tropical rainforests of the sea' is rooted in structural conditions that can be compared to terrestrial ecosystems of the tropics: i.e. growth rates are in tune with autochthonous species and their associated symbionts, ensuring a stable, long-term equilibrium, similar to tropical rain forests (Hubbell 1997). It is not surprising then that community structures of coral reefs easily cover a developmental time window stretching over thousands of years. The sudden emergence of modern reefs across the globe as we know them today was initiated some 6000 to 9000 years ago (Hallock 1997; Pandolfi 2002).

G. Witzany, *Biocommunication and Natural Genome Editing*,
DOI 10.1007/978-90-481-3319-2_4, © Springer Science+Business Media B.V. 2010

Over the last decades, however, the scientific community has noticed a drastic increase in coral-associated diseases of tropical reefs (Sutherland et al. 2004). Caribbean reefs are hardest-hit, earning this region the title 'coral disease hot spot' (Weil 2004).

In this review we will demonstrate that corals, being predominantly sessile organisms, are an essential part of the manifold communication patterns present within these ecosystems. Thus, internal communication patterns correspond and are coevolutionarily adapted to external signals of non-coral organisms. The coral animal, therefore, must be able to distinguish between self and non-self. According to their tolerance limits, these animals process and evaluate information in order to adjust their responses adequately. Equipped with appropriate recognition patterns, and in the absence of external interference, coral animals are able to thrive and establish themselves in oligotrophic environments of the tropics.

With regard to this highly diverse communicative competence we note that all this is possible only because communication processes within the animal (intraorganismic) and among the same and related species (inter-organismic) as well as species of non-related taxa (trans-organismic) are coordinated in an orderly manner, thereby enabling a context-coherent coordination of behavioural patterns. As long as these processes are comprehended, and interactions as well as exchange of information successfully established, the organism will prosper. Once these patterns become corrupted, via natural or anthropogenically-induced alterations, the coral animal faces an interactive uphill struggle. If external disturbances interfere with communication processes, as in the case of chronic events, it will compromise the animal's state of health and ultimately induce disease or even death. In order to understand better the survival strategies of corals it is necessary to work out the underlying interactional pattern – an issue that is of great relevance, especially when we consider the effects of global climate change (Hoegh-Guldberg et al. 2007).

We recognise that the use of signal molecules goes beyond information exchange, but above all requires the formation of various behavioural patterns necessary for successful interactions. The different forms of the symbiotic association, for example, with its endosymbionts and on the mucosa of the coral animal – here exosymbionts – stress the fact that conduct of symbiotic partners is beneficial for both parties as long as they are characterised by the absence of stress factors. As there are altruistic forms of interactions between coral and symbionts so there are defence response patterns, which can also involve life-and death situations (see below).

In any case, the context of a given situation determines the meaning of the used signs: (a) growth and (b) development are different modes of behaviour and need other patterns of signalling than (c) defence or (d) reproductive patterns. Likewise, (e) mutualistic symbioses require different forms of coordination from those of (f) commensalism or (e) parasitism. Thus, this systematic approach of coral communication demonstrates that the meaning (semantics) of signal molecules is context-dependent, and helps to give a better understanding of the full range of sign-mediated interactions of coral life.

4.2 Semiochemical Vocabulary of Corals

Communication patterns of signalling molecules within and among corals are amazingly complex. Depending on the developmental stage or other situational contexts, e.g. the reproductional process, different molecules are used (Fig. 4.1). Until now several classes of molecules which serve as signs in communicative processes has been identified:

- Secondary metabolities such as terpenes, cholesterol, wax ester diterpenoids;
- Neurotransmitter such as acetylcholine, dopamine, norepinephrine, serotonin;
- Hormones such as alloaromadendrine, noraromadendrine, cyclosinularane, prostanoids, terpenoids
- RNAs such as microRNAs and RNAi

Varying behavioural patterns lead to the production of different signals with different functions: antimicrobials, antifungals, corresponding secondary metabolites, and hormones (Kim 1994; Slattery et al. 1995, 1999; Ramesh and Venkateswarlu 1999; Iwashima et al. 2000; Kim et al. 2000a, b; Yasumoto et al. 2000; Roussis et al. 2001; Twan et al. 2003; Watanabe et al. 2003; Iguchi et al. 2004; Zhang et al. 2005).

As will be demonstrated later, corals possess a broad variety of hormonal substances for different behaviours, e.g. reproduction cycles and defence patterns

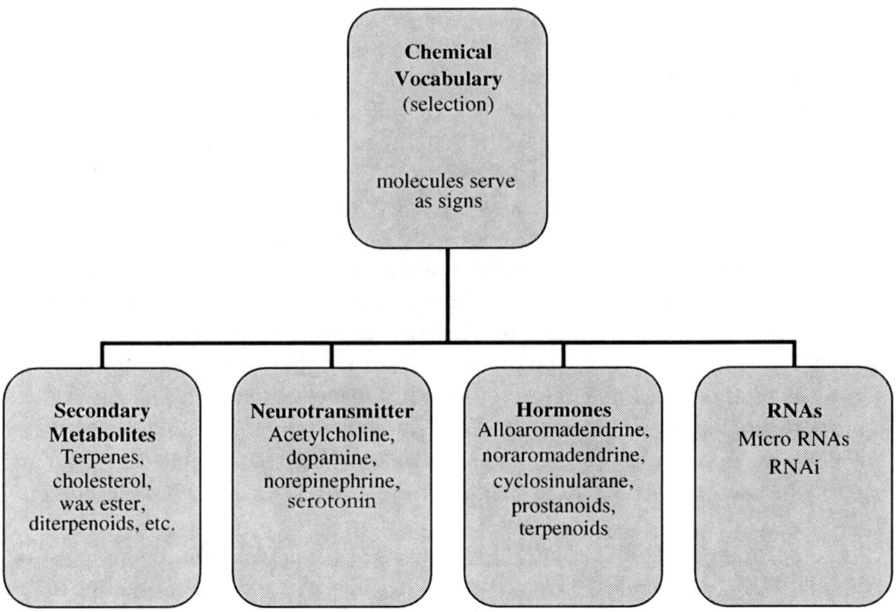

Fig. 4.1 Chemical vocabulary in coral communication processes

against opportunistic microbes, carnivores, herbivores and fungal infections (Hay et al. 1987, Kim et al. 2000a, b; Slattery et al. 1999).

4.3 Interpretation of External Influences

Abiotic interferences affect all eukaryotic organisms at any cellular level. Corals are particularly affected, as they are keystone species in tropical reef ecosystems (Bak et al. 1982; McClanahan and Maina 2003). Natural disasters such as earthquakes, cyclones and tsunamis change and thereby shape reef ecosystems. Under natural conditions such interference, results in change of species composition, favouring pioneering and faster-growing species, thereby initiating a new cycle of coral succession. Connell's intermediate-disturbance hypothesis (IDH) gives a good summary of the associated affects of such disturbances (Connell 1978; Begon et al. 1996; Loya 2004). Under moderate conditions, abiotic influences enable coral animals to release signalling chemicals via ocean currents (Harrison et al. 1984; Penland et al. 2004). Morphogenesis is yet another characteristic of stony corals that is likewise determined by abiotic conditions, i.e. light intensity, swell and surge patterns, oceanic currents (Geistner 1977; Horiguchi et al. 1999; Gleason et al. 2005; Stambler and Dubinsky 2005; Vargas-Angel et al. 2006; Veron and Stafford-Smith 2000).

In the context of anthropogenic influences, eutrophication owing to land-based intensive farming or mariculture, excessive fishing practices, and other man-made influences also interferes with coral-ecosystem communication, thereby easily tipping the balance towards a filter-feeding community. High phosphate levels in otherwise oligotrophic waters, for example, rapidly shut down the calcification mechanism (phosphate poisoning). In addition, higher nutrient levels favour (macro)-algae and prevent larval settling, with the result of a readily replacement of the coral-symbiont biotope (Tanner 1995; Loya 2004). Once filter feeders substitute for the coral carpet and become the dominant organisms, it is almost impossible to re-establish a thriving coral ecosystem within a reasonably short period (Hatcher 1997; Elmqvist et al. 2003). Closely associated with eutrophication and coastal degradation is the global trend in the decline of mangrove forest cover. Mangroves are an essential part of the wider tropical reef ecosystem, as their numerous prop roots account for a very rich and diverse habitat ranging from algae, sponges, and marine invertebrates to nurseries for young shrimp and coral fishes. Habitat degradation for shrimp farms and other coastal utilisation not only increases coastal erosion, but likewise interferes with the communication processes within the wider reef ecosystem, thereby detrimentally feeding back onto adjacent reefs (Mumby et al. 2004).

Another man-made disturbance originates from chemical interference; just one example is the pesticide Chlorpyrifos used on golf courses. Coral larvae are extremely sensitive to chemical signals at levels that are well below the detectable limits of current human technology. Since crustose coralline algae easily take up this pesticide, coral larvae have lower recruitment rates on substrates exposed to the chemical at concentrations even as low as 5 ppb than on untreated controls (Richmond 1997).

4.4 Transorganismic (Trans-Species) Communication

Both mechanical and chemical sign-mediated interactions of corals with other genera, families, and phyla, as well as with members of other kingdoms, are not only essential for their survival, but are the basis of coordination and organisation. These interactions cover the entire range – from the mutually supporting over neutral to harmful behaviours (Fig. 4.2). The variety of symbiotic communications, for example, requires very different behaviours from the partners involved (Weis et al. 2001).

Direct and indirect defence mechanisms are manifold and complement each other. Corals possess a 'non-self' warning system, especially when confronted with opportunistic microbes (Rohwer and Kelley 2004). Such microbes include single-celled autotrophic and heterotrophic prokaryotes, autotrophic and heterotrophic eukaryotes, as well as viruses. In fact, these opportunists can be found in all three domains of life on our planet, i.e. bacteria, archaea, and eukarya (Sherr and Sherr 2000).

Corals interact with non-related species through their muco-poly-saccharide layer (MPSL). This species-specific layer forms a boundary through which dissolved nutrients and gases diffuse. Hence, the mucus is a barrier against oppor-

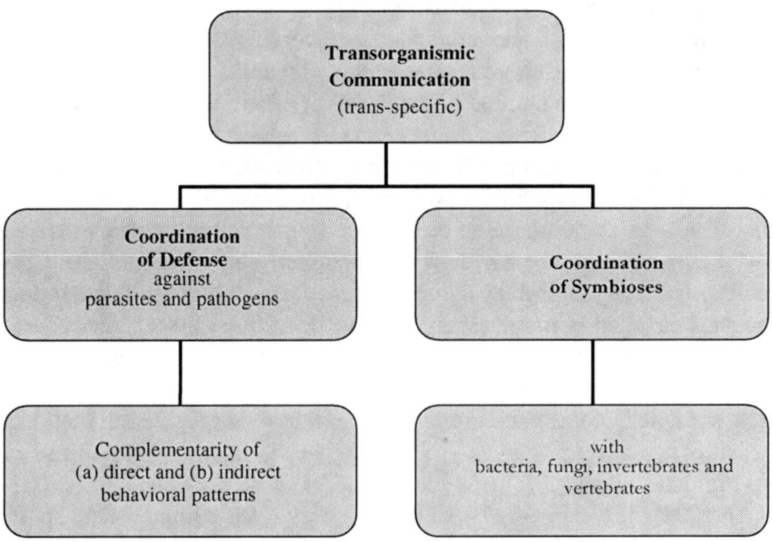

Fig. 4.2 Levels of transorganismic (trans-specific) communication

tunistic pathogens and should be considered the primary immune organ of corals. Beneficial prokaryotic residents living on and within the MPSL are yet another shield against opportunistic settlers (Shnit-Orland and Kushmaro 2009) and act as host-associated microbial community (Sunagawa et al. 2009). There are large numbers of mucus-adapted microbes, such as phosphate and nitrogen fixers. It is reported that even cold-water corals actively ‘fish’ the surface layer to obtain additional nutrients (Neulinger et al. 2008). In order to do so, corals encourage growth of specific microbes by the secretion of specialised mucus (Kushmaro and Kramarsky-Winter 2004). This in turn provides specialised microbiota as the ideal substrate to protect the coral animal from opportunistic settlers by occupying entry niches and/or through the formation of inhibition zones, e.g. prokaryotically-mediated production of antibiotics. Any disruption of the highly diversified microbial density on the MPSL will render the coral holobiont more susceptible to opportunistic pathogens, thereby enabling diseases to become established, and ultimately may result in the decline of the whole organism (Rohwer and Kelley 2004).

4.4.1 Coordination of Defence and Regeneration

Being predominantly sessile in lifestyle, corals are equipped with additional defence mechanisms against mechanically-induced damage and microbes (Gunthorpe and Cameron 1990; Kramarsky-Winter 2004). It appears that corals can differentiate between the various modes of inflicted tissue damage, i.e. triggered by (a) viruses, (b) bacteria, (c) fungi and (d) invertebrates as well as (e) damage induced by vertebrates. According to the type of damage, corals secrete different combinations of substances that serve to deal with such lesions.

A localised wound response leads not only to the production of mobile signal molecules, but requires a systematic reaction involving the entire organism – obviously, a mature colony has more reserves than a juvenile recruit (Alker et al. 2004). Reactions to lesions stimulate the transmigration of specific amoeboid wound cells toward the site of injury. The action of amoebocytes can be considered as first-aid measures, as they clean up cellular debris. Necrotic tissue is then sloughed off. Only now will the surrounding tissues stretch to cover as much of the wound as possible. If lesions are too large to be covered by stretching tissues, a much slower process takes over: tissue regrowth over the denuded area. It was observed, however, that corals affected by microbial-induced disease revealed a significant reduction of amoebocytes engaged in tissue repair (Kramarsky-Winter 2004). Amoebocytes are also involved in coral responses to pathogenic and temperature stress (Mydlarz et al. 2008).

Corals are able to coordinate directly and indirectly their complementing protective measures to varying degrees – according to the intensity of damage inflicted by opportunistic species (Chadwick-Furman and Rinkevich 1994; Koh 1997; Rinkevich 2004). The exposure to parasites or pathogens stimulates the coral organism to produce a specific array of immune substances (Bigger and Olano 1993; Rinkevich et al. 1994; Hildemann et al. 1977).

In addition, corals also produce enzymes that render their tissues unpalatable to certain predators (Lindquist and Hay 1996; Kelman et al. 1999). Contrary to the predictable succession of developmental phases of individual organisms, modular organisms can proliferate at one end while at the other tissues may be already in the phase of decomposition. Death in such organisms often results from becoming too big or succumbing to disease rather than from programmed senescence. Thus, the body of a modular organism has an age structure – it is composed of young and developing, actively functioning, as well as senescent, parts (Begon et al. 1996; Vytopil and Willis 2001). The modular structure enables corals to respond adequately to spatial limitations, predators, and unfavourable environmental conditions. Their morphology, in particular branching species, not only provides shelter for juvenile fish species and other invertebrates, but actively benefits the survivorship of the coral: e.g. the crab Trapezia sp. can even deter such vicious corallivorous predators as Acanthaster planci (Gosliner et al. 1996; Pratchett 2001).

Although corals are almost at the trophic base of shallow tropical reef biota, they are not passive ‘prisoners’ of their environment. Rather the opposite is the case: being at the base, or more appropriately at the centre, of what a reef community is all about, they actively shape their environment and thereby provide other organisms from other taxa with both habitat and substrate.

The fragility of reef-ecosystems becomes evident, however, when external disturbances such as devastating events, e.g. storms or changes in water quality, result in large deviations from normal modal flux rates that can lead to rapid reef decline. When limited in duration, as briefly mentioned in the context of Connell’s IDH, such events can stimulate another cycle of reef accretion (Hatcher 1997). In healthy reef ecosystems, this balance of production and bioerosion of reef biomass and calcium-carbonate substrate is closely balanced with net accumulation barely ahead of net reef loss. It regulates and influences ecosystem function of growth and decay (Glynn 1997; Paulay 1997). Under the influence of global climate change, however, nonlinear effects of long-lasting adverse conditions can easily lead to a situation where a ‘catch-’ or ‘keep up’ reef can easily precipitate into a ‘give-up’ state, where ‘permanent’ decline sets in (Adey 1978).

4.4.2 Communicative Coordination of Symbioses

Generally, there are multiple symbiotic interactions, which can be clearly differentiated into mutualistic, aggressive and defensive properties (Van Veghel et al. 1996; Hay 1997). Corals serve as symbiotic hosts for algae, protists, fungi, and a variety of bacterial and archaeal communities as well as viruses (Marhaver et al. 2008). Owing to the gradual increase in global sea-surface temperatures, communication processes between endosymbionts and coral hosts are increasingly disturbed (Baird et al. 2009; Rosenberg et al. 2009). In such cases, the coral’s ability to neutralise endosymbiotic production of radicals is compromised (Lesser 2004). Hence thermal stress combined with high irradiance pushes the host into the distress phase which leads to ever more frequent expulsion of endosymbiotic algae as temperature extremes exceed thermal threshold levels (Hoegh-Guldberg 1999, 2004). During extended bleaching events coral communities of entire reef sections lose their ability to regenerate and ultimately fade out, giving rise to a completely altered ecosystem (Edmunds and Gates 2003; Rowan 2004; Jones et al. 2008). There are, however, some clades of Symbiodinum that are better adapted to the higher temperatures and therefore develop different symbiotic interactions (Stat et al. 2008). Under certain circumstances corals are able to swap less temperature-tolerant clades for better-adapted ones (Buddemeier et al. 2004; Sampayo et al. 2008), thereby enabling corals to partly regain their vitality (Rowan 2004). The most marked differences among clades of Symbiodinium can be found between corals of the Caribbean and those in the Indo-Pacific region, while lesser differences exist among clades along a depth gradient within a given region (Toller et al. 2001).

Tropical reef corals are ecologically important examples of mutually helpful associations, whose success depends on the intrinsically interwoven webs of life. This network of complex communicative interactions shapes a sensitive dynamic equilibrium. Yet, as illustrated previously, this balance can far too easily be disturbed by outside influences, thereby triggering the collapse of entire reef sections. Besides active conservation efforts, only an exact analysis of the communication patterns of all the partners involved can aid a better understanding of coral physiology, disease patterns, and biotic and abiotic stress-factors (Edmunds and Gates 2003).

4.5 Interorganismic (Species-Specific and Species-Related) Communication

Species-specific and species-related sign-mediated interactions are termed interorganismic communication. Hermatypic reef corals are social organisms and with few exceptions, such as some members of Fungiidae, they are predominantly colonial and modular in appearance. This social capacity implies species-specific sign-mediated interactions, which enables corals of the same or a similar species, as well as distantly related relatives, to coordinate their behaviour. This coordination is most obvious during reproduction.

While some corals reproduce sexually, which requires synchronisation of opposite sexes, such as in cases of mass spawning, others do so asexually via budding or fragmentation. Most corals employ both modes of reproduction (Miller and Ayre 2004). Gametogenesis is a time- and resource-dependent process, so each method depends on favourable environmental conditions and developmental stages. Laboratory breeding experiments with some Acropora species have shown that pre-mating barriers are not strong. This indicates that chemical messengers are almost absent or not recognised in some of these species, in order to prevent merger of gametes from different species. The effect of such hybridisation can range from sterile individuals to introgression, speciation and even extinction (Wallace 1999). Hence it comes as no surprise that species involved in mass spawning overcome this limitation by expelling their gametes at precisely different time- windows of the lunar cycle coupled with solar insulation, and/or sea-surface temperatures (Harrison et al. 1984; Penland et al. 2004; Twan et al. 2006).

Evolutionary processes are also induced via hybridisation (Márquez et al. 2002; Miller and Van Oppen 2003). In situ observations along the GBR revealed that in a single night up to 150 species of the highly cross-fertile genus Acropora spawn within hours of each other. High cross-fertilisation rates were documented in vivo, while molecular tree topologies confirmed non-monophyletic patterns. These bore little similarity to cladistic analysis based on skeletal morphology or to the fossil record, leading to the conclusion that hybridisation does contribute to the enormous success of these species (Ryan 2006; Van Oppen et al. 2001). Indeed, hybridisation and polyploidy are recognised not only in other coral genera but also among a wide variety of marine invertebrates (Pfenninger et al. 2002). Similarly, recombinational crosses between directly and indirectly developing species of sea urchin have resulted in novel gene expression (Nielsen et al. 2000) and novel ontogenetic pathways (Raff et al. 1999), leading to the creation of new larval morphologies (Ryan 2006). According to Williamson (2003), interphyletic hybridisation of marine invertebrates must be a major transfer route of marine larval forms between taxa through a wide variety of hybridisations (Ryan 2006; Combosch et al. 2008).

Corals can differentiate between 'self' and 'non-self' (Rinkevich et al. 1994; Rinkevich and Sakai 2001). In order to protect their own growth range against proliferating opportunists, corals take defensive measures against 'non-self' tissues (Fig. 4.3). Allelopathic reactions, such as the production of chemicals to signal presence and to limit excessive proliferation of neighbouring coral species, occur in very complex ways and in various combinations and gradations (Yamazato and Yeemin 1986). There are some completely different and complementary defence mechanisms, e.g. escape by growth, aggressive behaviour, allelochemicals and aggregation (Bruno and Witman 1996). Aggressive and defensive behavioural patterns reciprocally depend on the extent of physical contact (Bak et al. 1982; Ferriz-Dominguez

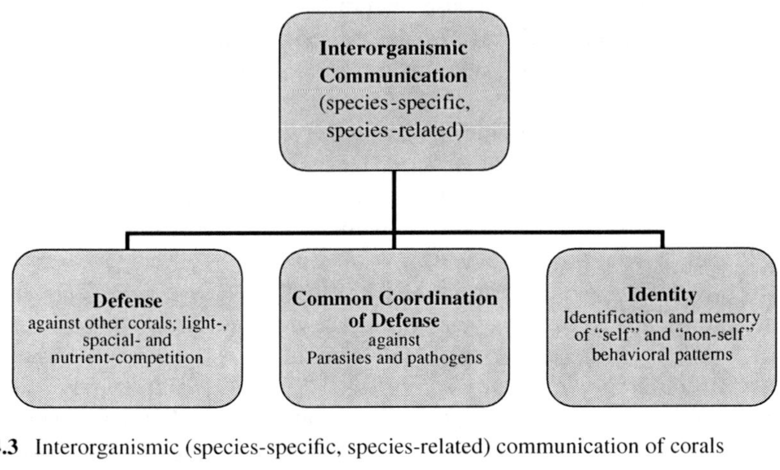

Fig. 4.3 Interorganismic (species-specific, species-related) communication of corals

and Horta-Puga 2001). Such responses can be quite broad, in that they can involve complete rejection of other coral colonies (use of sweeper and or stinging tentacles) or in the opposite case can result in complete merger of both colonies (Connell 1976; Cope 1982; Chadwick-Furman and Rinkevich 1994). Research employing various juvenile coral species showed that there are three kinds of response patterns: fusion, non-fusion and incompatible fusion. In the case of incompatible fusion, the junction of merging tissues lacks endosymbiotic algae. Slow-growing polyps characterise such an interfacial region. Over prolonged periods of time and as a result of this incompatibility, a skeletal barrier forms (Veron 1986; Hidaka et al. 1997).

Encounters of different coral species quite often result in subduing of the succumbing species by a dominant species. Overgrowth is simply one strategy to overcome spatial restriction in a flourishing reef community (Veron 1986; Frank and Rinkevich 2001).

Spatial and nutritional competition among corals and the concomitant stress factor do affect their fitness (Tanner 1997; Idjadi and Karlson 2007). Once resource limitations and habitat constraints are encountered, corals can become very aggressive and even kill competitors. It is well-known that sweeper tentacles, extrusions from the polyp's gut – up to five times the length of the polyp's tentacle and equipped with batteries of nematocysts – literally digest competitors. These extended mesenterial filaments are part of the macroscopic defence capabilities (Barnes and Hughes 1999). In addition, some corals generate considerable amounts of mucus loaded with nematocysts, which spreads out and over into the nearby environment to harm neighbouring colonies. Prolonged mucus production can significantly damage and even kill affected colonies. Other corals kill via the excretion of chemical poisons into the adjacent water body, while others again secrete substances which render larval settlement of potential competitors unfavourable (Lang 1970, 1971, 1973; Logan 1984, 1986; Lang and Chornesky 1990; Geffen and Rosenberg 2005).

Once a coral detects the presence of a nearby species with its ability for non-self recognition, it reacts with the production of finely gradated antibacterial/cytotoxic substances that will be used against potential intruders, or even against related coral species. This is an acquired response pattern and becomes engrained as a result of continuous stimulation. Corals are able to learn, in that they compare a given stimulation pattern with bodily stored stimulation patterns of the past (Hildemann et al. 1977; Rinkevich 2004).

Hence, corals are capable of differentiating between tissues and chemicals of kin species and those of 'non-related' species. This is essential in order to avoid repercussions that negatively affect the survival rate of nearby individuals of the same species. Fitted with such tools of discrimination, corals are even able to differentiate the sex of their opponents (Ates 1989).

Along with their symbiotic partners, each colony must have some kind of sphere of individuality in order to survive and prosper. Once intruders compromise these preconditions, substances are produced and released into the water body that hamper growth and proliferation of nearby competitors (Kim 1994; Wilsanand et al. 1999; Kim et al. 2000a, b; Roussis et al. 2001; Slattery et al. 1995). For the same purpose, even sturdy coral species can produce and release fast-acting antibacterial agents into the environment (Geffen and Rosenberg 2005).

4.6 Intraorganismic Communication

As we will see, intra-organismic communication processes are sign-mediated interactions within cells (intracellular) and between other cells of the same organisms (intercellular). Both communication processes are of crucial importance for the coordination of growth and development. Similarly to multi-cellular organisms the exchange of information must be assured on a local scale as well as on a spatially larger scale. Only by responding as a whole and in a coherent fashion can coral organisms react to the corresponding developmental challenges and physiological influences.

4.6.1 Intercellular Communication

Intercellular communication serves to communicate events within coral tissues or compartments to remotely located cells or tissues (Fig. 4.4). An injured coral, for example, organises an integrated molecular, biochemical and cell-biological response. This also includes immunological reactions (Bigger and Olano 1993; Hildemann et al. 1977; Rinkevich 2004).

The status information about the coral's current state is constantly monitored. This kind of information is to some extent suppressed, however, when the organism undergoes periods of growth. Doing so enables expression of the pre-processed intermediate steps necessary to accommodate such phase transitions. Embryonic development in particular and its subsequent transition into the larval stage as well as the metamorphosis to a juvenile polyp require finely-tuned coordination of growth and development (Okubo and Motokawa 2007). Special signalling pathways initiate these steps. Neuropeptides, for example, are hormone-like substances that coordinate metamorphosis (Iwao et al. 2002).

Acquisition of environmental parameters is part of the coral's lifecycle. This kind of information is even passed on to later generations. This pattern becomes most evident during asexual reproduction when offspring respond by establishing an extensive carpet of juvenile thickets, while others show preferences for high irradiance, and yet others again cope well with shadier conditions (Bak and Engel 1979).

4.6.2 Intracellular Communication

Intracellular merger of symbiotic dinoflagellates with marine cnidarians is the most important prerequisite to ensure a highly productive and diverse reef ecosystem (Santos et al. 2002; Takabayashia et al. 2004). To assert that the effects of symbiosis are long-lasting, this endocytotic process is coordinated by the ApRab5 gene. If, however, expression of this gene is disturbed, it leads to sudden separation and expulsion of the symbionts (Chen et al. 2004; 2005).

Therefore, successful intracellular communication (Fig. 4.4) with the symbiogenetically between the symbiogenetically assimilated unicellular eukaryotic cells

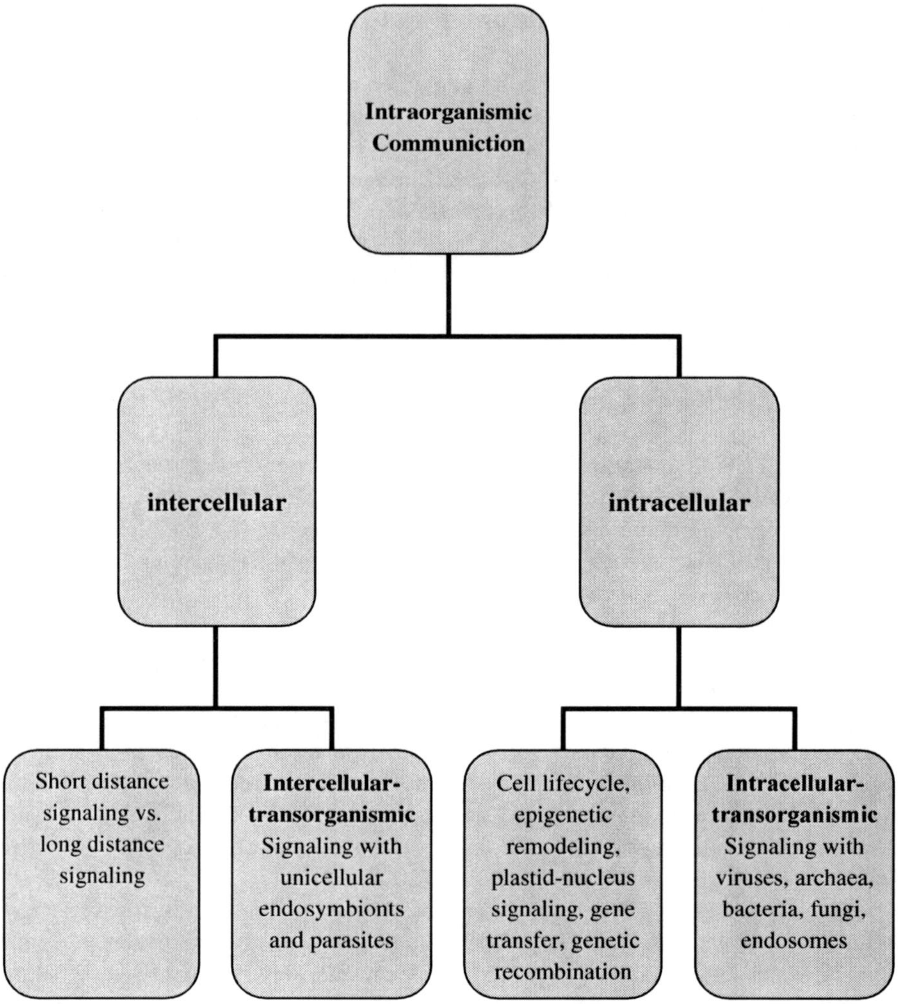

Fig. 4.4 Levels of intra-organismic communication

must take place. It makes sure that external messages are transformed and forwarded to the endsoymbionts. This information epigenetically influences the genetic expression of algal DNA. This in turn triggers a particular genetic reply, which leads to the production of signal molecules and generates an adequate response behaviour (Chen et al. 2000).

It seems quite obvious that viroids, viruses and bacteria do interfere in this kind of intracellular communication. Under extreme circumstances, this interference can disturb or even cause the collapse of the coral host – as in the case of induced tissue bleaching by Vibrio shiloi in Oculina patagonica or Vibrio coralliilyticus in Pocillopora damicornis (Rosenberg 2004; Rozenblat and Rosenberg 2004). Indeed most Symbiodinium species are infected with icosahedral double-stranded DNA-containing viruses. Under normal conditions they replicate without harming the host. Nonetheless, their latent virulence turns lytic (lethal to Symbiodinium) once water temperatures rise, thereby forcing the coral to expel their death endosymbionts, which also leads to tissue bleaching (Villarreal 2005). The elimination of the symbiont from the host tissue can vary between exocytosis, host cell detachment and host cell aptoptosis (Weis 2008).

Microbial interactions on the other hand are reciprocal; this enables incorporation of specific genetic features into the intruder's genome as well as the export of microbial datasets into those of the host organism. It is very likely that the ability to incorporate different traits in each other is a key principle of symbiogenetic processes (Shackelton and Holmes 2004). On a macroscopic scale, corals possess a decentralised neural network (Westfall and Sayyar 1997). Alignment of this network is never static, but implies neuronal-like plasticity. According to the animal's past experiences, it is the capacity to modify (increase or decrease) the magnitude of their connections. Both memory-functions and long-lasting neuronal plasticity require new RNAs for the appropriate protein synthesis. Such neuronal plasticity implies that signals are relayed via the synapse to the nucleus. Therein, these signals are converted in order to evoke a change in gene transcription (Fig. 4.4). Only then can the resulting changes (RNAs, proteins) be converted and sent back to the synapse to enable long-lasting change (Moccia et al. 2003; Martin 2004; Thompson et al. 2004). Also in this case it would be interesting to observe investigative efforts that tackle the decoding functions responsible for intracellular communication processes within corals.

Variation in pigmentation among corals is primarily the result of a few genes and their associated proteins in the endosymbionts. It is worth noting that the various colour patterns are the result of phenotypic plasticity rather than species diversity, as previously thought (Kelmanson and Matz 2003).

The simple organismic structure of Cnidaria phylum groups them almost into the 'archaic' section of the animal line as they share most characteristics with a common ancestor as well as those of modern animals of higher taxa, i.e. large-scale coral-algal-sponge bioherms emerged some 450 million years ago (Hallock 1997). It is worth mentioning that evolution, growth and development of the most important coral endosymbionts, i.e. members of the dinoflagellate genus Symbiodinium, assign to them a similar age. In this respect and based on the emergence of the first cnidarian precursors, the coral animal can be regarded as a 'window into the past' (Margulis and Schwartz 1988). Despite their simplicity, these animals have nearly as many genes as higher animal taxa – including vertebrates. Genome analysis has shown that most genes in Cnidaria can also be found in higher animals. This suggests that selective gene loss, or restriction of gene expression, rather than massive gene enrichment must be the main trigger for the emergence and evolution of higher animal taxa. This also implies that a selective decrease of the genes contributed to the sheer variety of animal species and paved the way for differentiation of the current known diversity in families, genera and species (Miller 2005).

4.7 Conclusion

There is compelling evidence that evolution, growth and development of scleractinian corals largely depend on successful trans-, inter- and intra-organismic communication processes. Indeed, it is not the individual coral organism that accounts for prosperous long-term establishment within the wider coral reef ecosystem, but rather the relationship and exchange of information with its surrounding that enables the coral to survive in the long run. Only once these processes are successfully established can coordination and organisation within the coral animal and with other corals take place in a controlled manner (Fig. 4.5). These processes enable corals to proliferate along with other organisms within the tropical reef ecosystem.

Communication processes should not be considered a mere exchange of information; rather they can be substantiated as diverse and manifold sign-mediated patterns of interaction. The use of such signs is subject to semiotic rules (Witzany 2006). The underlying rules are quite reliable and conservative.

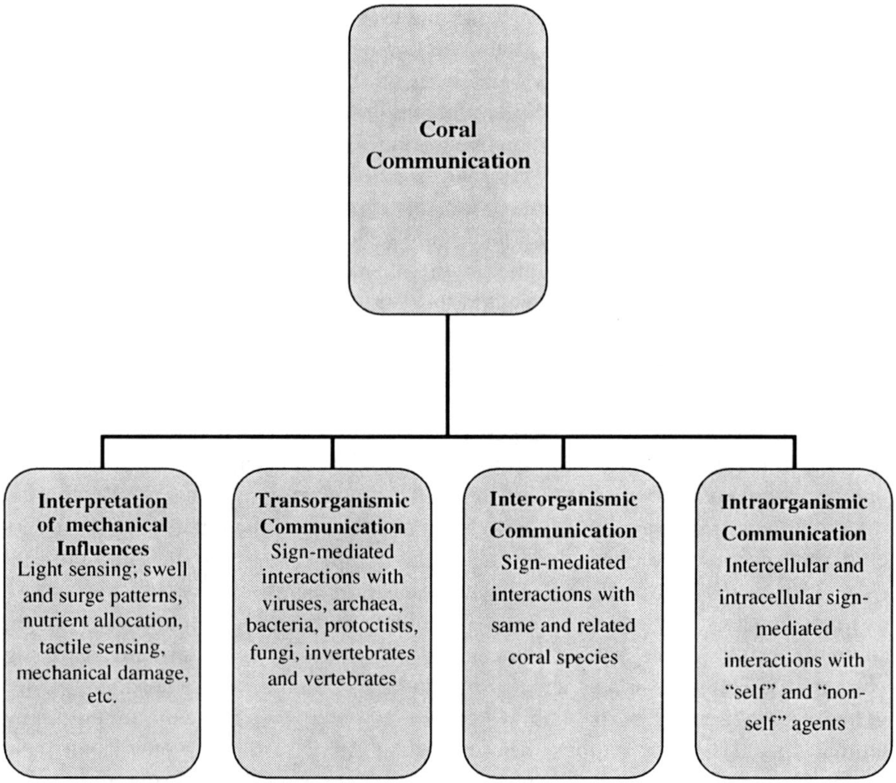

Fig. 4.5 Coral communication: sign-mediated interactions which each obey three levels of semiotic rules (syntactic, pragmatic, semantic)

Proliferation of corals depends on successful communication, which means the communication processes may even fail. The response of a particular interaction can be misleading, e.g. interaction between the coral and its endo- and exosymbionts pretends to be mutualistic, only to draw a temporary advantage from a given interaction and/or even to damage the exploited organism substantially once a shift in environmental conditions takes place. It is obvious that this cannot be the general form of communication. If all symbionts were to behave in this way, no individual species could survive in the long run. In the majority of cases, there must be efficient and successful interactions that are beneficial or at least not harmful for all the participants involved. Anything else leads to unsuccessful communication, disease, and decay of the coral organism. Hence, large-scale inefficiency in communication inevitably leads to long-lasting decline of the entire reef ecosystem. From this perspective we suggest that a beneficial behaviour exists among corals that serves to warn neighbouring organisms of the same species, as has already been observed in many other sessile taxa (Kobayashi et al. 2006).

Successful communication processes are guaranteed once all participants obey the semiotic rules in signal production and combination as well as transduction and interpretation-induced response behaviour. This does not imply, however, that these rules are devoid of errors or distortions. Nor does it mean that rules are always respected. In contrast with natural laws, the rules of sign-usage are in principle subject to change. Violation of this kind leads to distorted messages, which induce missing or even erratic response behaviours. There are various ways in which a message can be misinterpreted: (a) the sender incorrectly uses signs, they are mutilated and thereby miss the recipient, with the result that it hinders the receiver to respond to them appropriately; (b) the signs continuously encode messages which do not match with reality, e.g. messages indicating the settlement of opportunistic bacteria on the MPSL, increase in sea-surface temperature, etc; the receiver will become aware of the inappropriate content and over extended periods of time will ignore it and eventually adapt to this reality; (c) the sign continuously expresses a message which encodes for something else rather than what it is normally used for, e.g. to benefit unilaterally from a given interaction. In the context of chronic violation of the underlying rules, however, the organisation of life-processes is no longer possible, i.e. communicative coordination of evolution, reproduction, growth and development, in and among organisms, is chronically disturbed.

If communication processes in trans-, inter- and intra-organismic levels can be discerned in the future, then the various levels of rules in which signs are used can be differentiated. It is obvious that the 'molecular syntax' (Witzany 1995) of intracellular sign-usage is different from that of intercellular usage. Similarly, syntactic rules, which determine combinations of signalling molecules, differ in species-specific interactions from those used in trans-specific interactions. Embedded in the pragmatic level, i.e. in the ecological contexts and according to the behavioral contexts, different semantics are used. That is why a structurally identical message attains a partially different meaning when used in another context (Witzany 2005a, b; Bresgen 2007). In the future, the integration of this perspective could help to decipher coral-specific meanings of semiochemical messages in their entirety. The more we understand these communication processes and the meaning of the used signs, the deeper our understanding of the entire symbiology of corals will be.

References

Adey WH (1978) Coral reef morphogenesis: a multidimensional model. Science 202:831–837
Alker AP, Kim K, Dube DH et al. (2004) Localized induction of a generalized response against multiple biotic agents in Caribbean sea fans. Coral Reefs 23:397–405
Ates R (1989) Aggressive behavior in corals. FAMA 12:104–112
Baird AH, Bhagooli R, Ralph PJ et al. (2009) Coral bleaching. The role of the host. Trends Ecol Evol 24:16–20
Bak RPM, Engel MS (1979) Distribution, abundance and survival of juvenile hermatypic corals (Scleractinia) and the importance of life history strategies in the parent coral community. Mar Biol 54:341–352
Bak RPM, Termaat RM, Dekker R (1982) Complexity of coral interactions; influence of time, location of interaction and epifauna. Mar Biol 69:215–222
Barnes R, Hughes R (1999) An Introduction to Marine Ecology. 3rd ed. Blackwell Science, Oxford
Begon M, Harper JL, Townsend CR (1996) Ecology, Individuals, Populations and Communities. 3rd ed. Blackwell Science, Oxford
Bigger CH, Olano CT (1993) Alloimmune cellular responses of the gorgonian coral Swiftia exserta. J Immunol 150:134.
Bresgen N (2007) Signal and context. In: Witzany G (ed) Biosemiotics in Transdisciplinary Contexts. Proceedings of the Gathering in Biosemiotics 6, Salzburg 2006. Umweb, Helsinki
Bruno JF, Witman JD (1996) Defense mechanisms of scleractinian cup corals against overgrowth by colonial invertebrates. J Exp Mar Biol Ecol 207:229–241
Buddemeier R, Baker A, Fautin DG et al. (2004) The adaptive hypothesis of bleaching. In: Rosenberg E, Loya Y (eds) Coral Health and Disease. Springer Verlag, Berlin/Heidelberg
Chadwick-Furman N, Rinkevich B (1994) A complex allorecognition system in a reef-building coral: delayed responses, reversals and nontransitive hierarchies. Coral Reefs 13:57–63
Chen MC, Cheng YM, Hong MC et al. (2004) Molecular cloning of Rab5 (ApRab5) in Aiptasia pulchella and its retention in phagosomes harboring live zooxanthellae. Biochem Bioph Res Co 324:1024–1033
Chen MC, Hong MC, Huang YS et al. (2005) ApRab11, a cnidarian homologue of the recycling regulatory protein Rab11, is involved in the establishment and maintenance of the Aiptasia-Symbiodinium endosymbiosis. Biochem Bioph Res Co 338:1607–1616
Chen CA, Wallace CC, Yu JK et al. (2000) Strategies for amplification by polymerase chain reaction of the complete sequence of the gene encoding nuclear large subunit ribosomal RNA in corals. Mar Biotechnol 2:558–570
Combosch DJ, Guzman HM, Schuhmacher H et al. (2008) Interspecific hybridization and restricted trans-Pacific gene flow in the Tropical Eastern Pacific Pocillopora. Mol Ecol 17:1304–1312
Connell JH (1976) Competitive interactions and the species diversity of corals. In: Mackie GO (ed) Coelenterate Ecology and Behavior. Plenum Press, New York
Connell JH (1978) Diversity in tropical rainforests and coral reefs. Science 1999:1302–1310
Cope M (1982) Interspecific coral interactions in Hong Kong. Proc 4th Int Coral Reef Symp Manila 2:357–362
Edmunds PJ, Gates RD (2003) Has coral bleaching delayed our understanding of fundamental aspects of coral–dinoflagellate symbioses? BioScience 53:976–980
Elmqvist T, Folke C, Nyström M et al. (2003) Response diversity, ecosystem change, and resilience. Front Ecol Environ 1:488–494
Ferriz-Dominguez N, Horta-Puga G (2001) Short-term aggressive behavior in scleractinian corals from La Blan-quilla reef, Veracruz Reef System. Rev Biol Trop 49:67–75

Frank U, Rinkevich B (2001) Alloimmune memory is absent in the Red Sea hydrocoral Millepora dichotoma. J Exp Zool 291:25–39

Geffen Y, Rosenberg E (2005) Stress-induced rapid release of antibacterials by scleractinian corals. Mar Biol 146:931–935

Geistner J (1977) The influence of wave exposure on the ecological zonation of Caribbean Coral Reefs. Proc 3rd Int Coral Reef Symp Miami 2:23–29

Gleason DF, Edmunds PJ, Gates RD (2005) Ultraviolet radiation effects on the behavior and recruitment of larvae from the reef coral Porites astreoides. Mar Biol 148:503–512

Glynn PW (1997) Bioerosion and coral-reef growth: a dynamic balance. In: Birkeland C (ed). Life and Death of Coral Reefs. Chapman and Hall, London

Gosliner TM, Behrens DW, Williams GC (1996) Coral Reef Animals of the Indo-Pacific – Animal life from Africa to Hawaii Exclusive of the Vertebrates. Sea Challengers, Monterey

Gunthorpe L, Cameron AM (1990) Toxic exudates from the hard coral Goniopora tenuidens. Toxicon 28:1347–1350

Hallock P (1997) Reefs and reef limestones in earth history. In: Birkeland C (ed) Life and Death of Coral Reefs. Chapman and Hall, London

Harrison PL, Babcock RC, Bull GD et al. (1984) Mass spawning in tropical reef corals. Science 223:1186–1189

Hatcher BG (1997) Organic production and decomposition. In: Birkeland C (ed) Life and Death of Coral Reefs. Chapman and Hall, London

Hay ME (1997) The ecology and evolution of seaweed-herbivore interactions on coral reefs. Coral Reefs 16:67–76

Hay ME, Fenical W, Gustafson K (1987) Chemical defense against diverse coral-reef herbivores. Ecology 68:1581–1591

Hidaka M, Yurugi K, Sunagawa S et al. (1997) Contact reactions between young colonies of the coral Pocillopora damicornis. Coral Reefs 16:13–20

Hildemann WH, Raison RL, Cheung G et al. (1977) Immunological specificity and memory in a scleractinian coral. Nature 270:219–223.

Hoegh-Guldberg O (1999) Climate change, coral bleaching and the future of the world's coral reefs. Mar Freshwater Res 50:839–866

Hoegh-Guldberg O (2004) Coral reefs and projections of future change. In: Rosenberg E, Loya Y (eds) Coral Health and Disease. Springer Verlag, Berlin/Heidelberg

Hoegh-Guldberg O, Mumby PJ, Hooten AJ et al. (2007) Coral reefs under rapid climate change and ocean acidification. Science 318:1737–1742

Horiguchi T, Kawai H, Kubota M et al. (1999) Phototactic responses of four marine dinoflagellates with different types of eyespot and chloroplast. Phycol Res 47:101–107

Hubbell SP (1997) A unified theory of biogeography and relative species abundance and its application to tropical rain forests and coral reefs. Coral Reefs 16:9–21

Idjadi JA, Karlson RH (2007) Spatial arrangement of competitors influences coexistence of reef-building corals. Ecology 88:2449–2454

Iguchi K, Kukaya T, Yasumoto A et al. (2004) New marine sesquiterpenoids and diterpenoids from the Okinawan soft coral Clavularia koellikeri. J Nat Prod 67:577–583

Iwao K, Fujisawa T, Hatta M (2002) A cnidarian neuropeptide of the GLW-amide family induces metamorphosis of reef-building corals in the genus Acropora. Coral Reefs 21:127–129

Iwashima M, Nara K, Iguchi K (2000) New marine steroids, yonarasterols, isolated from the okinawan soft coral, Clavularia viridis. Steroids 65:130–137

Jones AM, Berkelmans R, van Oppen MJ et al. (2008) A community change in the algal endosymbionts of a scleractinian coral following a natural bleaching event: field evidence of acclimatization. Proc Biol Sci 275:1359–1365

Kelman D, Benayahu Y, Kashman Y (1999) Chemical defence of the soft coral Parerythropodium fulvum fulvum (Forskal) in the Red Sea against generalist reef fish. J Exp Mar Biol Ecol 238:127–137

Kelmanson IV, Matz MV (2003) Molecular basis and evolutionary origins of color diversity in great star coral montastraea cavernosa (Scleractinia: Faviida). Mol Biol Evol 20:1125–1133

Kim K (1994) Antimicrobial activity in gorgonian corals (Coelenterata, Octocorallia). Coral Reefs 13:75–80
Kim K, Harvell CD, Kim PD et al. (2000b) Fungal disease resistance of Caribbean sea fan corals (Gorgonia spp.). Mar Biol 136:259–267
Kim K, Kim PD, Alker AP et al. (2000a) Chemical resistance of gorgonian corals against fungal infections. Mar Biol 137:393–401
Kobayashi Y, Yamamura N, Sabelis MW (2006) Evolution of talking plants in tritrophic context: Conditions for uninfested plants to attract predators prior to herbivore attack. J Theor Biol 243:361–374
Koh EGL (1997) Do scleractinian corals engage in chemical warfare against microbes? J Chem Ecol 23:379–398
Kramarsky-Winter E (2004) What can regeneration process tell us about coral disease? In: Rosenberg E, Loya Y (eds) Coral Health and Disease. Springer Verlag, Berlin/Heidelberg
Kushmaro A, Kramarsky-Winter E (2004) Bacteria as a source of coral nutrition. In: Rosenberg E, Loya Y (eds) Coral Health and Disease. Springer Verlag, Berlin/Heidelberg
Lang JC (1970) Inter-specific aggression within the scleractinian reef corals. Ph.D. thesis, Yale Univ, New Haven
Lang JC (1971) Interspecific aggression by scleractinian corals; 1. The rediscovery of Scolymia cubensis (Milne-Edwards and Haime). B Mar Sci 21:952–959
Lang JC (1973) Interspecific aggression by scleractinian corals. Why the race is not only to the swift. B Mar Sci 23:260–279
Lang JC, Chornesky EA (1990) Competition between scleractinian reef corals- A review of mechanisms and effects. In: Dubinsky Z (ed) Coral Reefs; Ecosystems of the World 25. Elsevier, Amsterdam
Lesser MP (2004) Experimental biology of coral reef ecosystems. J Exp Mar Biol Ecol 300: 217–252
Lindquist N, Hay ME (1996) Palatability and chemical defense of marine invertebrate larvae. Ecol Monogr 66:431–450
Logan A (1984) Interspecific aggression in hermatypic corals from Bermuda. Coral Reefs 3: 131–138
Logan A (1986) Aggressive behavior in reef corals: a strategy for survival. Sea Frontiers 32: 347–351
Loya Y (2004) The coral reefs of eilat – past, present and future. In: Rosenberg E, Loya Y (eds) Coral Health and Disease. Springer Verlag, Berlin/Heidelberg
Margulis L, Schwartz KV (1988) Five Kingdoms. W. H. Freeman and Company, New York
Marhaver KL, Edwards RA, Rohwer F (2008) Viral communities associated with healthy and bleaching corals. Environ Microbiol 10:2277–2286
Martin KC (2004) Local protein synthesis during axon guidance and synaptic plasticity. Curr Opin Neurobiol 14:305–310
McClanahan TR, Maina J (2003) Response of coral assemblages to the interaction between natural temperature variation and rare warm-water events. Ecosystems 6:551–563
Miller DJ (2005) Lessons from coral genomics - perspectives on the development and evolution of higher animals. Biologisches Kolloquium der Christian-Albrechts-Universität, Kiel.
Miller KJ, Ayre DJ (2004) The role of sexual and asexual reproduction in structuring high latitude populations of the reef coral Pocillopora damicornis. Heredity 92:557–568
Miller DJ, van Oppen MJH (2003) A ‘fair go’ for coral hybridisation. Mol Ecol 12: 805–807
Moccia R, Chen D, Lyles V et al. (2003) An unbiased cDNA library prepared from isolated Aplysia sensory neuron processes is enriched for cytoskeletal and translational mRNAs. J Neurosci 23:9409–9417
Mumby PJ, Edwards AJ, Arias-Gonzales JE et al. (2004) Mangroves enhance the biomass of coral reef fish communities in the Caribbean. Nature 427:533–536
Mydlarz LD, Holthouse SF, Peters EC et al. (2008) Cellular responses in sea fan corals: granular amoebocytes react to pathogen and climate stressors. PLoS ONE doi:10.1371/journal.pone.0001811

Márquez LM, van Oppen MJH, Willis BL et al. (2002) Sympatric populations of the highly cross-fertile coral species Acropora hyacinthus and Acropora cytherea are genetically distinct. Proc Natl Acad Sci USA 269:1289–1294

Neulinger SC, Järnegren J, Ludvigsen M et al. (2008) Phenotype-specific bacterial communities in the cold-water coral Lophelia pertusa (Scleractinia) and their implications for the coral's nutrition, health, and distribution. Appl Environ Microbiol 74:7272–7285

Nielsen MG, Wilson KA, Raff EC et al. (2000) Novel gene expression patterns in hybrid embryos between species with different modes of development. Evol Dev 2:133–144

Okubo N, Motokawa T (2007) Embryogenesis in the reef-building coral Acropora spp. Zoolog Sci 24:1169–1177

Pandolfi J (2002) Coral community dynamics at multiple scales. Coral Reefs 21:13–23

Paulay G (1997) Diversity and distribution of reef organisms. In: Birkeland C (ed) Life and Death of Coral Reefs. Chapman and Hall, London

Penland L, Loulechad J, Idip D et al. (2004) Coral spawning in the western Pacific Ocean is related to solar insolation: evidence of multiple spawning events in Palau. Coral Reefs 23:133–140

Pfenninger M, Reinhardt F, Streit B (2002) Evidence for cryptic hybridisation between different evolutionary lineages of invasive clam genus Corbicula (Veneroida, Bivalvia). J Evol Biol 15:818–829

Pratchett MS (2001) Influence of coral symbionts on feeding preferences of crown-of-thorns starfish Acanthaster planci in the western Pacific. Mar Ecol Prog Ser 214:111–119

Raff EC, Popod EM, Sly BJ et al. (1999) A novel ontogenetic pathway in hybrid embryos between species with different modes of development. Development 126:1937–1945

Ramesh P, Venkateswarlu Y (1999) Novel steroid constituents of the soft coral Sinularia dissecta. Steroids 64:785–789

Richmond RH (1997) Reproduction and recruitment in corals: critical links in the persistence of reefs. In: Birkeland C (ed) Life and Death of Coral Reefs. Chapman and Hall, London

Rinkevich B (2004) Allorecognition and xenorecognition in reef corals: a decade of interactions. Hydrobiologia 530/531:443–450

Rinkevich B, Frank U, Bak RPM et al. (1994) Alloimmune responses between Acropora hemprichi conspecifics: nontransitive patterns of overgrowth and delayed cytotoxicity. Mar Biol 118: 731–737

Rinkevich B, Sakai K (2001) Interspecific interactions among species of the coral genus Porites from Okinawa. Jpn J Zool 104:91–97

Rohwer F, Kelley S (2004) Culture-independent analyses of coral-associated microbes. In: Rosenberg E, Loya Y (eds) Coral Health and Disease. Springer Verlag, Berlin/Heidelberg

Rosenberg E (2004) The bacterial disease hypothesis of coral bleaching. In: Rosenberg E, Loya Y (eds) Coral Health and Disease. Springer Verlag, Berlin/Heidelberg

Rosenberg E, Kushmaro A, Kramarsky-Winter E et al. (2009) The role of microorganisms in coral bleaching. ISME J 3:139–146

Roussis V, Chinou IB, Tsitsimpikou C et al. (2001) Antibacterial secondary metabolites Antibacterial activity of volatile secondary metabolites from Caribbean soft corals of the genus Gorgonia. Flavour Frag J 16:364–366

Rowan R (2004) Coral bleaching: Thermal adaptation in reef coral symbionts. Nature 430:742

Rozenblat YBH, Rosenberg E (2004) Temperature-regulated bleaching and tissue lysis of pocillopora damicornis by the novel pathogen Vibrio coralliilyticus. In: Rosenberg E, Loya Y (eds) Coral Health and Disease. Springer Verlag, Berlin/Heidelberg

Ryan FP (2006) Genomic creativity and natural selection: a modern synthesis. Biol J Linn Soc 88:655–672

Sampayo EM, Ridgway T, Bongaerts P et al. (2008) Bleaching susceptibility and mortality of corals are determined by fine-scale differences in symbiont type. Proc Natl Acad Sci USA 105:10444–10449

Santos SR, Taylor DJ, Kinzie A et al. (2002) Molecular phylogeny of symbiotic dinoflagellates inferred from partial chloroplast large subunit (23S)-rDNA sequences. Mol Phylogenet Evol 23:97–111

Shackelton LA, Holmes EC (2004) The evolution of large DNA viruses: combining genomic information of viruses and their hosts. Trends Microbiol 12:458–465
Sherr E, Sherr B (2000) Marine microbes – an overview. In: Kirchman DL, Williams PJB (eds) Microbial Ecology of the Oceans. Willey-Liss Inc, New York
Shnit-Orland M, Kushmaro A (2009) Coral mucus-asociated bacteria: a possible first line of defense. FEMS Microbiol Ecol 67:371–380
Slattery M, Hines GA, Starmer J et al. (1999) Chemical signals in gametogenesis, spawning, and larval settlement and defense of the soft coral Sinularia polydactyla. Coral Reefs 18:75–84
Slattery M, James A, McClintock JB, Heine JN (1995) Chemical defenses in Antarctic soft corals: evidence for antifouling compounds. J Exp Mar Biol Ecol 190:61–77
Stambler N, Dubinsky Z (2005) Corals as light collectors: an integrating sphere approach. Coral Reefs 24:1–9.
Stat M, Morris E, Gates RD (2008) Functional diversity in coral– dinoflagellate symbiosis. Proc Natl Acad Sci USA 105:9256–9261
Sunagawa S, Desantis DZ, Piceno YM et al. (2009) Bacterial diversity and white plague disease-associated community changes in the Carribean coral montatstraea faveolata. ISME J doi: 10.1038/ismej.2008.131
Sutherland K, Porter JW, Torres C (2004) Disease and immunity in Caribbean and Indo-Pacific zooxanthellate corals. Mar Ecol Prog Ser 266:273–302
Takabayashia M, Santos CR, Cook CB (2004) Mitochondrial DNA phylogeny of the symbiotic dinoflagellates (Symbiodinium, Dinophyta). J Phycol 40:160–164
Tanner JE (1995) Competition between scleractinian corals and macroalgae: an experimental investigation of coral growth, survival and reproduction. J Exp Mar Biol Ecol 190: 151–168
Tanner JE (1997) Interspecific competition reduces fitness in scleractinian corals. J Exp Mar Biol Ecol 214:19–34
Thompson KR, Otis KO, Chen DY et al. (2004) Synapse to nucleus signaling during long-term synaptic plasticity: a role for the classical active nuclear import pathway. Neuron 44:997–1009
Toller WW, Rowan R, Knowlton N (2001) Zooxanthellae of montastraea annularis species complex: patterns of distrubution of four taxa of symbiodinium on different reefs and across depths. Biol Bull 201:348–359
Twan WH, Hwang JS, Chang CF (2003) Sex steroids in scleratinian coral, euphyllia ancora: implication in mass spawning. Biol Repro D 68:2255–2260
Twan WH, Hwang JS, Lee YH et al. (2006) Hormones and reproduction in scleractinian corals. Comp Biochem Physiol A Mol Integr Physiol 144:247–253
Van Oppen JH, McDonald BJ, Willis B et al. (2001) The evolutionary history of the coral genus Acropora (Scleractinia, Cnidaria) based on a mitochondrial and a nuclear marker: reticulation, incomplete lineage sorting, or morphological convergence? Mol Biol Evol 18:1315–1329
Van Veghel MLJ, Cleary DFR, Bak RPM (1996) Interspecific interactions and competitive ability of the polymorphic reef-building coral Montastrea annularis. B Mar Sci 58:792–803
Vargas-Angel B, Colley SB, Hoke SM et al. (2006) The reproductive seasonality and gametogenic cycle of Acropora cervicornis off Broward County, Florida, USA. Coral Reefs 25:110–122
Veron JEN (1986) Corals of Australia and the Indo-Pacific. University of Hawaii Press, Honolulu
Veron JEN, Stafford-Smith M (2000) Corals of the World. Vol 3, Australian Institute of Marine Science, Townsville
Villarreal LP (2005) Viruses and the Evolution of Life. ASM Press, Washington
Vytopil E, Willis B (2001) Epifaunal community structure in Acropora spp. (Scleractinia) on the Great Barrier Reef: implications of coral morphology and habitat complexity. Coral Reefs 20:281–288
Wallace C (1999) Staghorn Corals of the World – A Revision of the Genus Acropora. CSIRO Publishing, Collingwood
Watanabe K, Sekine M, Iguchi K (2003) Isolation of three marine prostanoids, possible biosynthetic intermediates for clavulones, from the Okinawan soft coral Clavularia viridis. Pharm Bull 51:909–913

Weil E (2004) Coral reef diseases in the wider Caribbean. In: Rosenberg E, Loya Y (eds) Coral Health and Disease. Springer Verlag, Berlin/Heidelberg

Weis VM (2008) Cellular mechanisms of Cnidarian bleaching: stress causes the collapse of symbiosis. Exp Biol 211:3059–3066

Weis VM, Reynolds WS, deBoer MD et al. (2001) Host-symbiont specificity during onset of symbiosis between the dinoflagellates Symbiodinium spp. and planula larvae of the scleractinian coral Fungia scutaria. Coral Reefs 20:301–308

Westfall JA, Sayyar KL (1997) Ultrastructure of neurons and synapses in the tentacle epidermis of the sea an-emone calliactis parasitica. J Morphol 232: 207–216

Williamson DI (2003) The Origins of Larvae. Kluwer Academic Publishers, Dordrecht

Wilsanand V, Wagh AB, Bapuji M (1999) Antibacterial activities of anthozoan corals on some marine microfoulers. Microbios 99:137–145

Witzany G (1995) From the 'logic of the molecular syntax' to molecular pragmatism. Evol Cog 1:148–168

Witzany G (2005a) Natural history of life: history of communication logics and dynamics. SEED J 5:27–55

Witzany G (2005b) From biosphere to semiosphere to social lifeworld; biology as an understanding social science. tripleC 3:51–74

Witzany G (2006) The Logos of the Bios 1; Contributions to the Foundation of a Three-Leveled Biosemiotics. Umweb, Helsinki

Yamazato K, Yeemin T (1986) Preliminary study on the inter- and intraspecific interactions among corals of Khang Khao Island, the Sichang Islands, Gulf of Thailand. Galaxea 5:163–174

Yasumoto M, Mada K, Ooi T et al. (2000) New terpenoid components from the volatile oils of the soft corals Clavularia viridis and Sarcophyton acutangulum. J Nat Prod 63:1534–1536

Zhang GW, Ma XQ, Kurihara H et al. (2005) New hemiketal steroid from the soft coral Cladiella sp. Org Lett 7:991–994

Chapter 5
Biocommunication of Fungal Organisms

Abstract Fungal Organisms are the only eukaryotes which assemble unicellular and multicellular organisms which seems to be an indicator of their ancestral role in the evolution of multicellular eukaryotes such as animals and plants. Coordination and organisation processes occur in all organismic kingdoms, and in fungi these are seen during the formation of fruiting bodies (intraorganismic), between species of the same kind (interorganismic) and between non-fungal organisms (transorganismic). These involve rule-governed sign-mediated interactions, the signalling processes being nothing other than distinct biocommunicative processes. The semio-chemicals used are of biotic origin, in contrast to chemical indicators that trigger the fungal organism to react in a specific manner.

5.1 Introduction

Some people enjoy forest mushrooms, i.e. fruiting bodies of fungi, via manifold variations of delicious dishes. However, when fruit, vegetables or bread are coated with spore-building surfaces we know that they are no longer edible, as some of the poisons produced by these fungi are known carcinogens. The presence of fungi is even less appreciated when we are confronted by fungal skin infections – they are unpleasant and require laborious medical procedures. Fungal diseases in the oral cavity, throat or even in internal organs are not just unpleasant, they are dangerous and can even be life threatening, although some of them are closely connected with hominid evolution (Lott et al. 2006). In agriculture it is known that fungi are a common cause for recurring disease events on cultivated plants. Animal farms and maricultural activities also have to deal with fungal diseases.

Although fungi are not really striking organisms, it is estimated that there are at least 1.5 million species, out of which about 300 000 are described in the scientific literature.

Although it is known that fungal spores spread via the air, we hardly recognise their omnipresence, mostly because single spores cannot be seen by the naked eye. Being suspended in the air, they just wait to settle on a suitable substrate,

G. Witzany, *Biocommunication and Natural Genome Editing*,
DOI 10.1007/978-90-481-3319-2_5, © Springer Science+Business Media B.V. 2010

especially when temperature and moisture offer ideal conditions for activating the spores, germinating, initiating hyphal growth and developing mycelia.

On an evolutionary time scale, the kingdom of fungi emerged approx. 300 million years after the appearance of the first animal species, although they descended from a common ancestor (Lang et al. 2002). On top of that, they share common traits with eukaryotic organisms. In contrast to animal- and higher plant-life, monocellular representatives are fairly common among fungi, i.e., fungi are by no means mere multicellular organisms. This fact can be easily and coherently reconstructed through the lineages of protoctista – in that coordinated behavioural patterns are found among single-celled eukaryotes, which closely resemble those of single-celled fungi. However, there are unmistakable and significant differences in protoctist structure (i.e. flagellated) and those of fungi (i.e. non-flagellated). Obviously, fungi have evolved out of protoctists, such as red and joch-algae (Margulis and Schwartz 1988).

Fairly early associates of the fungal kingdom are those organisms that interact symbiotically with fungi, as is the case with lichens. All higher fungal life-forms originated from these symbiotic ways of life, which later became independent by detaching themselves from this close and vital dependence. Lichens are symbiotic partnerships. As pioneering organisms, they may settle on bare rock. Being essential bioeroders, they extract nutrients from mineral matter, thereby initiating the process of soil formation and paving the way for successive organisms that include root-forming plants.

As with plants, fungi are sessile organisms that can live for extremely long periods or extend over large areas: one example has been found which covers as much as 15 hectares with an age of approx. 1500 years. Endolithic fungi from the Antarctic are known to be among the most long-lived organisms on this planet (Villarreal 2005). Another example covers 900 hectares with an estimated age of 2400 years (Volk 2002).

Higher fungi are modular hyphal organisms in that they reproduce by clonation or are parasexual. They establish interlocking networks. Like red algae they merge their cytoplasms to form a multi-nucleated cell. This holistically-merged body is also found among some plants. A spore germinates under appropriate environmental conditions and is followed by the formation of filaments called hyphae. The latter are characterised by nuclear division and spore formation, which develop into monokaryotic filaments (tip-growing). As in other tip-growing organisms, this process requires a high density of tip-localised mitochondria (Levina and Lew 2006).

The embryological stage, a characteristic of higher plants and animals, is completely absent in fungi (Margulis and Schwartz 1988). Hyphae formation is also found among certain bacteria, such as Streptomyces and Actinomyces. When hyphae are tightly packed together into a mat this is called a mycelium. Each filament of hyphae has tubular side walls made mostly of chitin, a feature that is also common in Arthropoda. The cell walls that seem to separate adjacent cells in a filament are called septa – however, their porous nature does not really give them separating properties. The merger of filamental tips of the same or different species triggers a self/non-self identification process. This process is sign-mediated and results either in repulsion or attraction. If the latter occurs, merger into a dikaryotic mycelium takes place and initiates the formation of a fruiting body.

Most fungi are saprobes and decompose and feed on non-living organic matter. They secrete powerful enzymes that enable the cells to digest organic matter from the nearby environment outside their body, in turn breaking this down to smaller molecules that can be absorbed and reincorporated in a dissolved form. To deter potential predators, a number of complex and highly efficient deterring substances are produced by fungi (Margulis and Schwartz 1988; Ingold and Hudson 1993; Jennings 1995).

Fungi are known to utilise a broad variety of symbiotic interactions with animals, plants and eukaryotic protozoa for both mutual benefit as well as parasitic and even lethal associations. They also settle on specific types of tissue. Fungal diseases are known to affect both plant and animal life, in some cases inducing devastating effects on and also in organisms (Humber 2008), e.g. in agriculture. On the other hand, fungal activity has been actively exploited by humans for thousands of years in beer, wine, cheese and bread production. Since fungi are very simple organisms, sequencing and technical manipulation are relatively easy, making them ideal organisms for laboratory experiments (Weld et al. 2006), e.g. *Neurospora grassa* (Dunlap et al. 2004; FGSC 2008). Their application in the medical field as producers of various antibiotics is comprehensive and well-appreciated (e.g. penicillin).

5.2 Semiochemical Vocabulary of Fungi

So far five different primary signalling molecules have been shown to be involved in very different behavioural patterns such as filamentation, mating, growth, and pathogenicity (Adachi and Hamer 1998; Wang and Heitman 1999; Hemenway and Heitman 1999; Alspaugh et al. 2000; Borges-Walmsley and Walmsley 2000). Behavioural coordination and the production of such substances can only be achieved through interpretation and response behaviour. Furthermore, there are numerous, less-investigated subunits that play an accompanying role as they are weaker in effect (Lengeler et al. 2000). Globally, these semio-chemicals serve to coordinate similar goals in different fungal species, yet species-variation among them cannot be ignored.

The roles of these signalling molecules are as follows: (i) Mitogen activated protein kinase signalling (MAPK) is involved in cell integrity, cell wall construction, pheromones/mating and osmo-regulation (Dohlman and Slessareva 2006; Yu et al. 2008), (ii) the cyclic adenosine monophosphate cAMP/PKA system is involved in fungal development and virulence, (iii) the RAS protein is involved in cross-talk between signalling cascades, (iv) calcium-calmodulin-calcineurin are involved in cell survival under oxidative stress, high temperature, membrane/cell wall perturbation and (v) rapamycin is involved in the control of cell growth and proliferation (Fernandes et al. 2005) (Fig. 5.1).

To date, 400 different secondary metabolites have been documented. These are known to contain mycotoxins and are used both for defensive and aggressive behaviours.

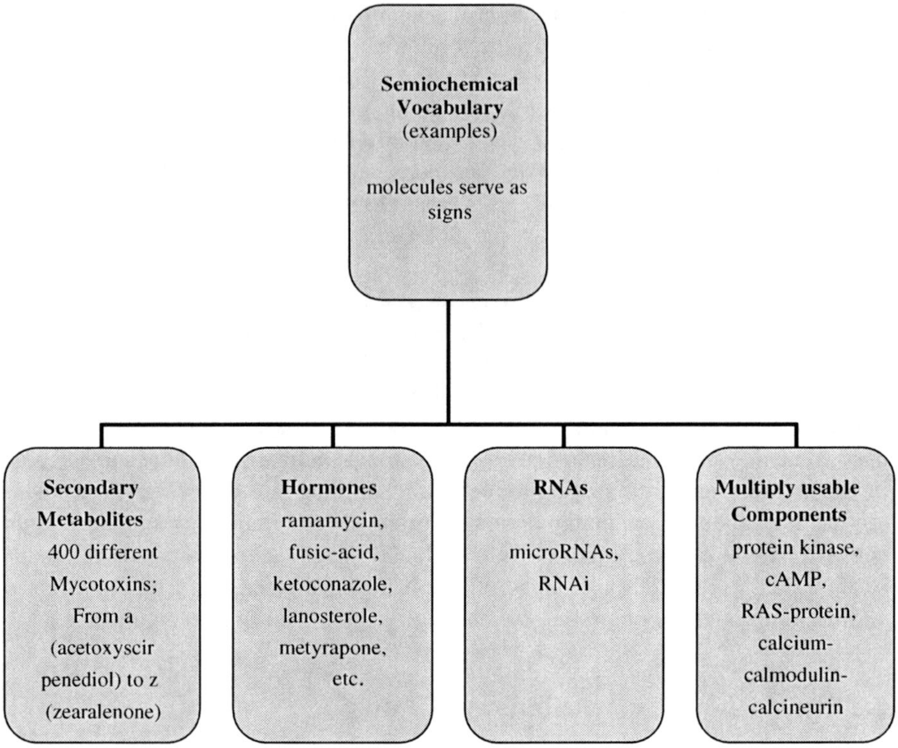

Fig. 5.1 Examples for semiochemicals in fungal communication processes

5.3 Interpretation of Abiotic Indices

Fungi react sensibly to varying nutrient availability and nutrient fluxes, and do so by responding holistically (via intra-organismic communication). For example, in the case of carbon or nitrogen insufficiencies, the internal communication system of the organism responds adequately and is phenotypically expressed by a change in hyphal growth. To date, two specific signalling pathways have been found that coordinate such behaviour. These diverging pathways have also been documented among other fungi, including those that are pathogenic to plants and animals (Lengeler et al. 2000).

An abundance of nutrients for example, results in increased production of cAMP (Lee et al. 2003), which suppresses that of Ste11, itself a mating inhibitor. Such processes are sign-mediated processes that originate from cyclase-associated proteins (CAP) (Lengeler et al. 2000). Nutrient availability has to be communicated to induce cell growth (Dechant and Peter 2008)

As with animals and plants, seasonality is found in fungi as a part of the circadian system (Dunlap et al. 2004; De Paula et al. 2008), e.g. light-regulated physiological processes that coordinate the internal fungal clock (Bell-Pederson et al. 1996).

Neurospora crassa, for example, responds according to irradiance patterns arising from the diurnal cycle. In such cases we can speak of a *Neurospora* chronotype (Dunlap et al. 2004; Tan et al. 2004). However, this responsiveness is not connected to photosynthetic activity, as this is only found in green algae, i.e. in the symbiotic association of lichens.

Fungi are capable to generate adaptive response behavioural patterns against environmental stress situations (Alonso-Monge et al. 2009) such as ROS reactive oxygen species (Moradas-Ferreira and Costa 2000) or in the case of varying osmotic conditions (Hohmann 2002; Santos and Shiozaki 2004; Krantz et al. 2006) which includes constant monitoring of cell wall integrity (Levin 2005).

Almost all fungi are saprobes and feed on dead organic matter. The excretion of extracellular digestive enzymes fragments larger biomolecules and makes them soluble, and more readily accessible for the fungal organism. This is particularly important for the digestion of cellulose, which is broken down by the enzymatic activity of exocellulase and endocellulase, and lignin (lignin peroxidase and manganese peroxidase). Enzymatic breakdown of organic matter yields simple sugars, amino acids, fatty acids and other smaller molecular components (Margulis and Schwartz 1988).

5.4 Transorganismic Communication

Fungi are competent to generate a great variety of signalling processes with non-fungal organisms such as plants, animals (insects) and bacteria. A typical example of this kind is the mutually beneficial symbiosis between bark beetles and quite a few different fungi (Sullivan and Berisford 2004). The fungal spores benefit from the locomotion provided by the beetles in several ways e.g. access to new hosts, while the beetle benefits from the availability of fungal nutrients and pheromones. Some fungi provide nitrogen, amino acids and sterols that are crucial for the development of beetle larvae – however, this takes effect only once the adult beetle has colonised a host. Many bark beetles have even evolved transportation pockets for fungal hyphae, a fact that points to their common evolutionary history (Kopper et al. 2004).

Another trans-organismic symbiotic signalling process happens between fungi and ants, and is derived from a co-evolutionary relationship that lasted over millions of years (Poulsen and Boomsma 2005). Interestingly some lignin-degrading fungi also produce semiochemicals that have effects on the feeding and foraging behaviour of a Formosan subterranean termite (Cornelius et al. 2004).

There is much evidence to suggest that the fungal and animal kingdom share common ancestors, such as protoctists, which have a true nucleus like choanoflagellates (Villarreal 2005). In this sense, fungi and animals are more related to each other than is the case with the plant kingdom. This is further strengthened by the sign-mediated processes, which regulate cellular functions. Yet another indicator of their common ancestry is found in a particular signalling pathway, termed the mitogen-activated protein kinase cascade (MAPK). This plays a crucial role in cell wall stabilisation of fungi and pheromone/mating interactions among mammalian cells. On top of that, MAPK is highly conserved (Lengeler et al. 2000).

Then there are fungi which parasitise plants. For example, they may colonise host tissues with an intercellular mycelium that forms haustoria, i.e. fungal mats within plant cells (Jakupovi et al. 2006) that penetrate the cell to utilise the nutrients of the plant. Studies on hazardous fungal infections on plants have revealed the crucial role of enzymes such as cutinase, pisatin, demethylase and HC-toxins (NIAID 1993) (Fig. 5.2).

Today, several hundred species of fungi colonise more than 100,000 different plant species. This type of cohabitation relies on and requires symbiotic signalling (Lammers 2004), e.g. the initiation of filamentous growth of fungi through plant hormones (Prusty et al. 2004).

Roots of plants provide better conditions for mycorrhizal fungi, which in turn supply plants with better nutrients (Brundrett 2002). For the fungus, such a relationship is either balanced or predatory. Endophytic fungi, however, live in plants without triggering symptoms of disease (Brundrett 2002). Today, scientists consider the origin of the plant cell to be the result of the terrestrial activity of mycorrhiza, i.e. settlement on land was a co-evolutionary event that is comparable to that between flowering plants and insects (Villarreal 2005). The mutually beneficial relationship between subterranean fungi and plant roots is a fine-tuned network of sign-mediated interactions developed over millions of years (Gross 2006), whereby fungi excrete digestive enzymes into the surrounding soil, and convert nutrients into aqueous solutions that in turn can be readily absorbed by the plant. A staggering 80% of all terrestrial plants rely on the activity of mycorrhiza, especially trees (Schwarze et al. 2004).

Fungi that affect animals are usually dependent on the host's body temperature, i.e. host colonisation by the fungi is only possible if the body temperature is sufficiently high. *Aspergillus fumigatus* in particular colonises animal hosts if they are

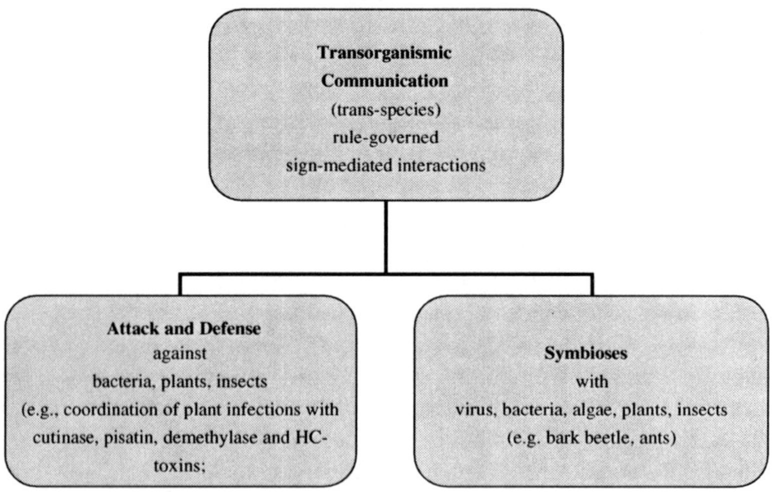

Fig. 5.2 Transorganismic (trans-species) fungal communication

under thermal stress (Bhabhra et al. 2006). Although fungal disease is common in birds, the relative resistance of endothermic vertebrates to fungal diseases may be a result of immune responses connected with higher body temperature (Casadevall 2006).

A typical fungus cultivated in research laboratories is *Neurospora*. Several subspecies of this genus contain toxins that are pathogenic for animals and plants. However, they are essential components for the large-scale industrial production of antibiotics, chemicals, enzymes and pharmaceuticals (Dunlap et al. 2004).

One of the most striking trans-organismic communication processes between fungi and non-fungal species can be found in lichens (Sanders 2001, 2006). Lichens are polyphyletic. They have been derived many times independently from different kinds of ascomycetes, so their nature of symbiosis doubtless varies (Raven 2001). In lichens, the algae provide the fungi with photosynthate, while the fungi provide the algae with nutrients. Lichens constitute one of the oldest known fungal members and are capable of resisting quite adverse environmental conditions. The symbiosis between fungi and algae or fungi and bacteria results in a mutual supply of nutrients and their associated competences. Through quorum sensing, the fungi benefit from the bacterial association (Hogan 2006). In turn, the bacteria utilise dissolved fungal metabolites to satisfy their nutritional requirements.

A similar co-dependence is observed with algae as symbiosis partners. But there is also a great variety of other sign-mediated interactions between fungi and bacteria that are beneficial for both and are based on reciprocal signalling (McAlester et al. 2008). Trans-organismic communication can be found in any interaction that involves fungi and viruses, bacteria, protoctista (algae), animals (insects) and plants.

5.5 Biocommunication Among Fungal Species

Since there are both single and multi-cellular species, the communication processes between the same species and related fungal species cannot be distinguished unambiguously from intercellular communication (intra-organismic). Thus communication processes in the mono-cellular yeast (Banuett 1998), which resemble that of amoeba like *Dictyostelium*, must be considered as inter-organismic communication, whereas those within multi-cellular fungal species are truly intercellular. To verify whether cell-to-cell communication is inter- or intra-organismic, we have to consider intercellular processes on a case-by-case basis.

Herein we find another fundamental characteristic of biota, concerning the competence for identifying 'self' and 'non-self' (Muirhead et al. 2002). This competence was successfully demonstrated in *Neurospora crassa* (Glass et al. 2000; Glass and Saupe 2002; Glass and Kaneko 2003). It is obvious that this capacity to distinguish between oneself and others is vital for fungi, since the encounter of mycelia from the same species results in the merger of their fungal hyphae. However, such dikaryotic mycelia can also result from the merger of different fungal species.

While peripheral hyphae tend to avoid merging with hyphae of other species, the opposite is the case with those at the centre of the mycelium (Glass et al. 2000;

Glass and Saupe 2002; Glass and Kaneko 2003). If one assigns mycelia the role of a wrapper within which the fungus is enveloped, so to speak, into a fluid-like continuum, then the nuclei of compatible but different species are 'flowing' through the same mycelium. The overall result is an organism, which houses nuclei of different genetic origin in its cytoplasm (Wu and Glass 2001). However, if specific genetic sequences are incompatible, then repulsion sets in, forcing the approaching hyphae into an immune-like response, much like those found in plants.

Resource competition in fungi occurs directly, indirectly and via mechanical interaction. Indirect competition involves absorption of all available resources within the reach of the mycelium, thereby starving potential competitors by maintaining a nutritional deficiency gradient. Direct interaction on the other hand, involves secondary metabolites, which suppress growth or even induce death of the competing fungi. The antibiotics employed in such cases can be either volatile or non-volatile. Mechanical interaction simply requires overgrowth of one fungal species by the other, in which the overgrowing species exerts its lytic action on the other. In some cases lysis is induced via antibiotic agents (Dix and Webster 1995; Griffin 1994).

Like bacteria, single-celled fungi also use quorum sensing to regulate and affect biofilm formation and pathogenesis (Reynolds and Fink 2001). This is mediated by small molecules that accumulate in the extra-cellular environment. If these reach a sufficiently high concentration a response regulator is activated within the local population of cells, leading to the coordination of gene expression (Hogan 2006).

In parasitic interactions between fungal organisms cytoplasmic fusion was found during infection processes, which indicated genetic transfer in the host-parasite relationship. The recognition pattern in this predator-prey relationship is mediated by trisporoids, which are also involved in a non-parasitic behavioural pattern, being responsible for sexual communication (Schultze et al. 2005) (Fig. 5.3).

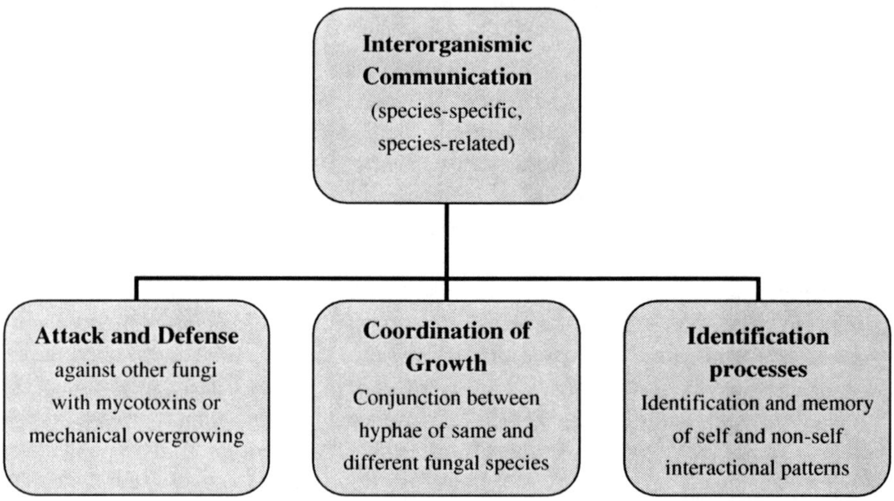

Fig. 5.3 Examples for behavioral contexts of interorganismic fungal communication

Some fungi have drawn attention only as saprobes, ectomycorrhizal symbionts or parasites of plants, while their role as parasites of other fungi has not yet been mentioned e.g. basidiomycetes. Their role in a great variety of interactions is still being studied (Bauer and Oberwinkler 2008).

5.6 Biocommunication Within Fungal Organisms

The countless variety of fungal organisms represents a major challenge when establishing a homogeneous designation of the signalling processes employed. Research activities so far have predominantly focused on those fungal species that pose a serious threat to agriculture, are pathogenic to humans or possess antibacterial properties. Species of this kind are relatively well investigated, whereas species with obviously insignificant properties are little known.

Growth in *Saccharomyces cerevisiae*, for example, is entirely dependent on the final purpose of growth, i.e. its growth and behavioural patterns are correlated. Thus mating of haploid fungal cells is a completely different behavioural cycle than filamentous growth of diploid cells. The signals involved can be differentiated by four different signal-related production patterns (Lengeler et al. 2000).

There is new evidence that certain proteins that have been restricted to signalling pathways in animals so far are also common in fungal signalling pathways (Herranz et al. 2005). In addition, apoptotic processes, which have been considered to play important roles in the development of multicellular organisms of animals and plants, have now been found to be important for regulating development, growth and aging processes in single-celled fungi (Hamann et al. 2008).

A very large family of transmembrane receptors is responsible for transducing extracellular signals into intracellular responses such as G protein-coupled receptors (Xue et al. 2008).

5.6.1 Intercellular Communication

Hyphal growth is a totally different behavioural pattern of conduct than normal cell growth: such cells change shape, become elongated, and re-orientate themselves into specific directions to physically come into contact or even merge with each other, only to colonise a potential growth resource.

Fungal hyphae simultaneously extend in a given direction only once nutritional resources are ideally distributed. However, this is a rare event. Usually the fungus propagates in the direction that is enriched with organic matter (carbon and its derivatives) just to halt growth once little or no resources are available; here the fungus coordinates its growth by establishing certain priorities. In order to do so, the fungus employs intercellular signals that enable it to comprehend the overall state of the organism. This includes an integrated ability to discriminate between self and non-self. Once the fungus encounters a resource-depleted substrate or even poisonous compounds, it responds by halting its growth cycle or by propagating in another direction. The protein signals involved in such processes are quite complex: the apex of hyphae houses specialised receptors that are able to respond to any environmental condition. Any carbon-enriched substrate causes these receptors to become active, which in turn results in the production and release of protein signals into the hyphal cytosol where the corresponding signalling cascade is triggered. In turn, the mycelium responds with the mobilisation and translocation of resources into the activated area. In the the absence of carbon-rich substrates, or at increasingly acidic pH levels, the hyphae respond by activating yet other receptors that slow down growth and eventually cause the organism to withdraw resources from the affected area. Since septa within the hyphae are perforated, they perform similar functions as the gap junctions in higher animals, micro-plasmodesmata in cyanobacteria and plasmodesmata in higher plants (Belozerskaya 1998; Gessler et al. 2007).

5.6.2 Intracellular Communication

Unfortunately, our understanding of fungal communication processes for the assembly and merger of cell walls, capsules, and cellular surface-receptors is incomplete. Little is known about the secreted signalling molecules that penetrate the cell walls. Even the phenomenon of endocytotic absorption of dissolved organic matter into the fungal cell is not really understood (NIAID 1993).

Yet there has been some progress in this field. By investigating a great number of signal transduction events from the outside through the cell membrane into the cytoplasm it has been possible to decode some important intracellular communication processes. Thus, it was found that signal processes coordinate cell polarity, mating, pheromone control and cellular morphology. Some of these processes even adjust the cell cycle, cause polarised growth activity and modify the transcription profiles of fungal cells (Lengeler et al. 2000; Bardwell 2004; Fernandes et al. 2005).

By examining the fungal pathogen *Paracoccidioides brasilienus* it was realised that some signalling pathways are identical to those of other species such as *Saccharomyces cerevisiae*, *Cryptococcus neoformans*, *Candida albicans* and *Aspergillus fumigatus* (Fernandes et al. 2005).

The 'protein cascades' that characterise production pathways of appropriate chemicals and messenger signals reflect the behavioural contexts, which are to some extent completely different (Fernandes et al. 2005):

- Cell integrity, cell wall construction, pheromone/mating, and osmo-regulation by mitogen-activated protein kinase signalling (MAPK)
- Fungal development and virulence by the cAMP/PKA system
- Cross-talk between cascades by the RAS protein
- Cell survival under oxidative stress, high temperature, membrane/cell wall perturbation by calcium-calmodulin-calcineurin
- Control of cell growth and proliferation by rapamycin

Combinatorial communication procedures such as those involving the MAPK and cAMP pathways are also part of the behavioural contexts (Lengeler et al. 2000). These in turn serve to multiply the semantic contents of the encoded messages.

Several signalling pathways have been found that sense extracellular stimuli and convert them into intracellular signals that regulate developmental and growth processes signalling. The guanine nucleotide-binding protein (G-Protein) is essential for extracellular detection of nutrients and sexual partners (Dohlman 2002; Kays and Borkovich 2004; Hoffman 2005). The TOR protein kinases, which are bound and inhibited by rapamycin, function as nutrient-sensing signals and regulate cellular responses like proliferation, transcription, translation, autophagy and ribosome biogenesis (Beck and Hall 1999; Cutler et al. 2001) (Fig. 5.4).

As with any signal-mediated interaction that can be achieved with molecules, the same components are employed for differing behavioural contexts and in varying messages. That is, different modes of behaviour can be achieved by syntactically identical signalling. The signalling pathways use identical proteins to coordinate

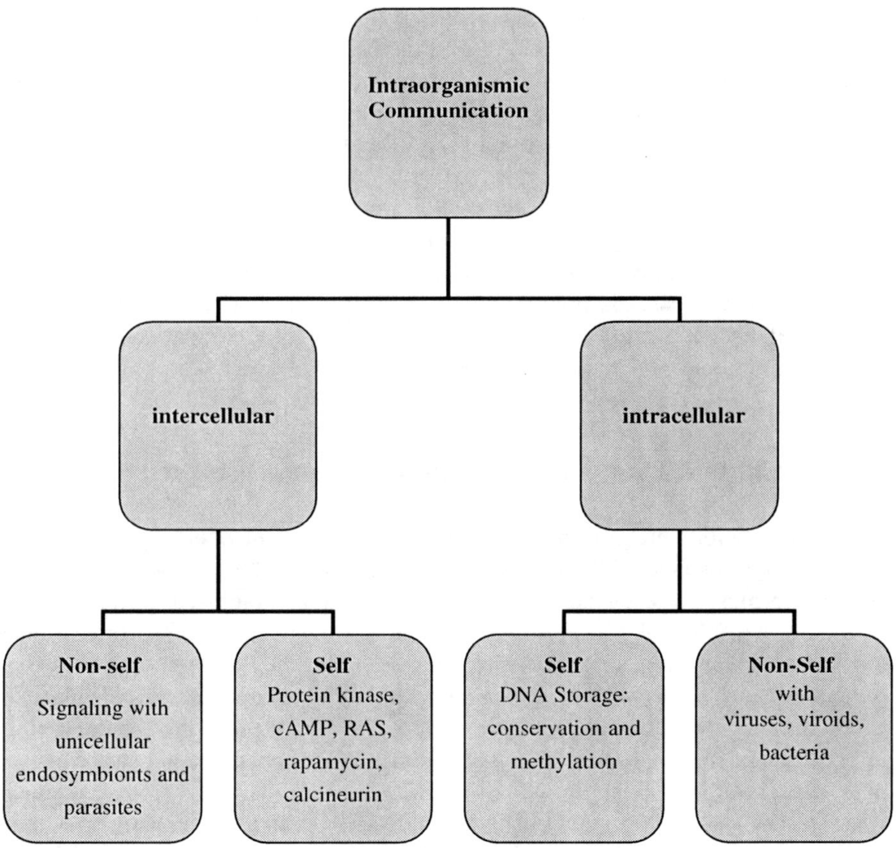

Fig. 5.4 Levels of intraorganismic fungal communication

different response patterns. Even if they are syntactically identical they semantically contain a completely different meaning: for example,

- activated cAMP triggers filamentous growth in *Saccharomyces cerevisiae*,
- regulates positive virulence in *Cryptococcus neoformans*,
- suppresses mating in *Schizosaccharomyces pombe* and
- inhibits filamentous growth in *Ustilago maydis* (Lengeler et al. 2000; D'Souza and Heitman 2001) or
- activates protein kinase for directly or indirectly induced developmental changes in *Magnaporthe grisea* during infection of rice (Mitchel and Dean 1995).

Another example is the *Ustilago maydis* pheromone response, which regulates both cell fusion and the pathogenicity programme for plant infection (Krüger et al. 1998; Hartmann et al. 1996, 1999).

These examples show that different behavioural contexts determine different meanings of identical signalling molecules. In such cases, biosemiotically identical signals can induce opposite responses in different organisms. It is interesting to note that fungi are not just capable of differentiating among varying messages and responding appropriately, but moreover are able to differentiate among molecules that are chemically identical to signalling molecules and obviously contain no relevant meaning ('noise'), i.e. are not parts of biotic messages.

Recent genome comparisons have provided new insights into the evolutionary aspects of fungi. The hypothesis that evolution happened by whole genome duplication events followed by selective gene-loss and stabilisation is strengthened by analysis on *Saccharomyces cerevisiae* (Kellis et al. 2004). Interestingly the signal-to-noise ratio in yeasts is approximately 70% protein coding regions and 15% regulatory elements in the non-protein coding regions, in comparison to humans with ratios of 3 and 97% respectively (Kellis et al. 2003).

5.6.3 Unique Relationship Between Fungi and Viruses

Fungal populations are commonly colonised by persistent double-stranded RNA viruses, single-stranded RNA viruses, double-stranded DNA viruses, retroviruses and even prions. They are the simplest organisms to be colonised by retroviruses, in the form of fungal retroposons. In contrast to all other eukaryotes, fungal mitochondria are infected by either double-stranded RNA or DNA viruses. Most fungus-infecting viruses are not pathogenic although they are associated with toxin genes that are pathologically dangerous, not for the infected host, but for non-infected fungal relatives and non-fungal organisms. This phenomenon has been used quite successfully by brewing industries, to protect useful yeasts from exogenous yeast genetic parasites by colonising the industrial yeast colonies with protective versions of killer viruses (Villarreal 2005). An example for mutually beneficial symbiosis is that of the black aspergillus species and dsRNA viruses, as in the case of mycovirus (van Diepeningen et al. 2006)

The interconnected filamentous fungi are a very attractive habitat for viruses because they allow rapid motility through the whole cellular network. Fungal networks also allow replacement of nuclei that are left behind in mitochondria. This may be an advantage if mitochondria are colonised by parasites, which is common (Villarreal 2005). It has been suggested that non-pathogenic viral colonisation of fungal mitochondria protects mtDNA from age-dependent degradation and is beneficial for the longevity of the fungal host. As a consequence, the growth radius of the fungal mats may increase up to 10-fold (Villarreal 2005).

Interestingly, higher fungi lack repeat elements in DNA, whilst lower fungi can have up to 50% of their genomes colonised by repeat elements. Recent research shows that these repeat elements represent remnants of early viral genomic colonisation events. Fungi were the first organisms to be colonised by viral RNA. This assumption is underlined by the presence of SINEs and LINEs. So far up to 15 SINE-families have been recognised (Whisson et al. 2005), a fact that supports the assumption that very early viral colonisation of the fungal genome must have taken place (Villarreal 2005; Rooney and Ward 2005).

Although plants and animals are descendants of fungi, the relationship between fungi and viruses is different from virus-animal or virus-plant relationships (Villarreal 2005). The latter two lack linear plasmids, killer phenotypes, mitochondrial infections, distorted senescence and ubiquitous double-stranded RNA colonisation, which are all characteristics of the fungal virus-host relationship. This may be an indication of co-evolutionary interactions between fungi and viruses (Villarreal 2005).

Genomic studies show that most methylated regions of *Neurospora crassa* are derived from repeat-induced point mutations (RIP), which comprise a pre-meiotic homology-based genome defense system. This early immune system consists of a variety of inactivated transposons that include DNA modifications and chromatin modifications (Selker et al. 2003; Galagan and Selker 2004). As shown by comparative genomics, such inactivated transposons descended from viral infection events that reached a persistent status, providing the host with a new phenotype of an innate immune system in order to prevent similar infections (Villarreal 2005).

In yeast species this transposon is derived from an endogenous retrovirus, i.e. the characteristic Ty1 element, which encodes a functional reverse transcriptase, and a gag gene, which encodes structural proteins. It is inserted into the silent regions of chromatin, which means that this is a good example of a beneficial situation for both the colonising virus and the host. Interestingly, yeasts also persistently harbour killer viruses that are lethal to related yeast species that are not infected persistently. Generally it is assumed that the high rate of different toxins in the organismic kingdom of fungi represents such toxin/antitoxin modules that are persistently integrated into the fungal host genomes (Villarreal 2009).

Fungal genomes are not highly populated by genetic settlers in comparison to animal or plant genomes. Only one family of endogenous retroviruses is present in several full-length copies, some of them completely deactivated in specific fungal lineages. These chromovirus sequences sometimes match with fungal host sequences. In Ascomycetes and Basidiomycetes these sequences are distinct but inactivated and not transcribed, which could be an indicator of several massive viral colonisation events during which the previously integrated viruses were displaced. Interestingly, a small number of viral copies are highly conserved and transcribed, so we can assume that they are closely connected with fungal evolution (Villarreal 2009). The high infection rate by dsRNA viruses and endogenous retroviruses is coherent with the presence of a small interfering RNA system with the ability for post-transcriptional silencing. This new system for identifying RNA and repeated DNA sequences by double-stranded RNA sequences is able to silence RNA expression, its main purpose being to silence endogenous retroviruses (chromovirus) and retroposons (Villarreal 2009)

Interestingly, the mitochondria of fungi are highly colonised by persistent double-stranded RNA viruses, each of them from another lineage but present in the host genome as multiple interconnected settlers. Especially the oxidative competence of mitochondrial respiration which is essential for life span of fungal host share some antagonist which regulates its function. If regulation of this process is disturbed, the life span of the host can be interrupted abruptly (Villarreal 2009).

Generally the unique relationship between fungi and viruses seems to be an indicator of the virus-driven evolutionary history of fungi in that the virus-host relationship, especially in the endogenous persistent status (represented by dsRNA and small linear dsDNA viruses), is linked to all immunity mechanisms in fungi, such as killer toxins in mitochondria or siRNA, i.e. dsRNA-based identification competence, with an important role in silencing endogenous retroviruses and limiting retroviral infections (Villarreal 2009). This would be coherent with the fact that most fungal genomes are small in comparison to other eukaryotic genomes that lack the ability to ward off RNA viruses and retroviruses.

5.7 Conclusion

An overview of all significant levels of fungal communication shows that identification of signal-mediated processes in signalling pathways are context dependent - both within and among fungal cells as well as between fungi and other organisms.

Such dependence is prevalent in both (beneficial or parasitic) colonisation and defense responses. Depending on the context, molecular components are integrated into unique signalling pathways where they attain the corresponding meanings. Such meanings are subject to change, i.e. they rely on various behavioural contexts, which differ under altered conditions. These contexts concern cell adhesion, pheromone response, calcium/calmodulin, cell integrity, osmotic growth, stress response or cell growth through rapamycin. The interactional context determines the semantic relationship, i.e., its meaning and the function of the chemical components, and forms a signal-mediated communication pattern in fungi. This is a common feature in all eukaryotic kingdoms: the context determines the meaning of trans-, inter- and intra-organismic (inter- and intracellular) communication, while differences in abiotic and biotic signal perception determine the content arrangement of response behaviour (Witzany 2006, 2007) (Fig. 5.5).

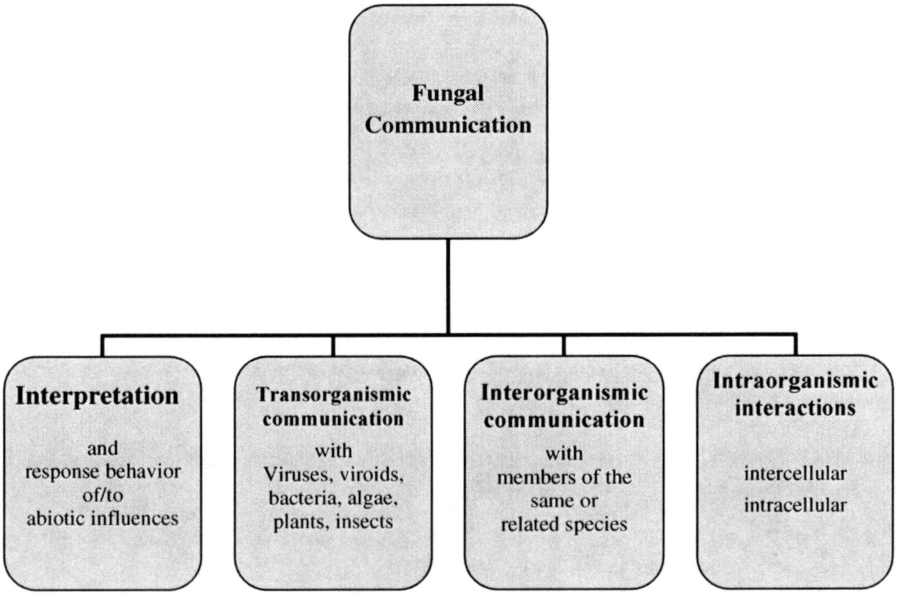

Fig. 5.5 General levels of fungal communication

After recognising how versatile fungal communication competences really are we can see that one main principle is followed throughout all these signalling processes: fungal organisms coordinate all their behavioural patterns with a core set of chemical molecules. The interactional context and the different modes of coordinating appropriate response behaviour in e.g. development, growth, mating, attack, defense, virulence, etc. determines the combinations of signals that generate the appropriate meaning-function, i.e. informational content of messages. These generating processes normally function in a very conservative way but under certain circumstances may fail, or selective pressure may lead to changes that can be a driving force in fungal evolution.

Additionally it can be recognised that the persistent lifestyle of viruses is a driving force in fungal evolution in that they are the main resource for immunity, group identity and a large number of important secondary metabolites.

References

Adachi K, Hamer JE (1998) Divergent cAMP signaling pathways regulate growth and pathogenesis in the rice blast fungus Magnaporthe grisea. Plant Cell 10:1361–1374

Alonso-Monge R, Román E, Arana DM, Pla J, Nombela C (2009) Fungi sensing environmental stress. Clin Microbiol Infect 15:17–19

Alspaugh JA, Cavallo LM, Perfect JR et al. (2000) Ras1 regulates filamenta-tion, mating and growth at high temperature of Cryptococcus neoformans. Mol Microbiol 36:352–365

Banuett F (1998) Signalling in the yeasts: an informational cascade with links to the filamentous fungi. Microbiol Mol Biol Rev 62:249–274

Bardwell L (2004) A walk-through of the yeast mating pheromone response pathway. Peptides 25:1465–1476
Bauer R, Oberwinkler F (2008) Cellular basidiomycete – fungus interactions. In: Varma A, Abbott L, Werner D et al. (eds) Plant Surface Microbiology. Springer, Berlin/Heidelberg
Beck T, Hall MN (1999) The TOR signalling pathway controls nuclear localization of nu-trient-regulated transcription factors. Nature 402:689–692
Bell-Pederson D, Dunlap JC, Loros JJ (1996) Distinct cis-acting elements mediate clock, light, and developmental regulation of the Neurospora crassa eas (ccg2)gene. Mol Cell Biol 16:513–521
Belozerskaya TA (1998) Cell-to-cell commu-nication in differentiation of mycelial fungi. Membr Cell Biol 11:831–840
Bhabhra R, Zhao W, Rhodes JC et al. (2006) Nucleolar localization of asper-gillus fumigatus CgrA is temperature-dependent. Fungal Genet Biol 43:1–64
Borges-Walmsley MI, Walmsley AR (2000) cAMP signalling in pathogenic fungi: control of dimorphic switching and pathogenicity. Trends Microbiol 8:133–141
Brundrett MC (2002) Coevolution of roots and mycorrhizas of land plants. New Phytol 154: 275–304
Casadevall A (2006) Fungal virulence, ver-tebrate endothermy, and dinosaur extinction: is there a connection? Fungal Genet Biol 42:98–106
Cornelius ML, Bland JM, Daigle DJ et al. (2004) Effect of a lignin-degrading fungus on feeding preferences of Formosan sub-terranean Termite (Isoptera: Rhinotermitidae) for different commercial lumber. J Econ Entomol 97:1025–1035
Cutler NS, Pan X, Heitman J et al. (2001) The TOR signal transduction cascade controls cellular differentiation in response to nutrients. Mol Biol Cell 12:4103–4113
De Paula RM, Lamb TM, Bennett L et al. (2008) A connection between MAPK pathways and circadian clocks. Cell Cycle 7:2630–2634
Dechant R, Peter M (2008) Nutrient signals driving cell growth. Curr Opin Cell Biol 20:678–687
Dix NJ, Webster J (1995) Fungal Ecology. Chapman & Hall, London
Dohlman HG (2002) G proteins and pheromone signalling. Ann Rev Physiol 64:129–152
Dohlman HG, Slessareva JE (2006) Pheromone signaling pathways in yeast. Sci Signal 364: cm6
Dunlap JC, Loros JJ, Denault D et al. (2004) Genetics and molecular biology of circadian rhythms. In: Brambl R, Marzluf GA (eds) The Mycota III. Biochemistry and Molecular Biology. 2nd ed. Springer-Verlag, Berlin/Heidelberg
D'Souza CA, Heitman J (2001) Conserved cAMP signaling cascades regulate fungal de-velopment and virulence. FEMS Microbiol Rev 25:349–364
FGSC- Fungal Genetics Stock Center (2005). The Neurospora Homepage. http://www.fgsc.net/Neurospora/neurospora.html
Fernandes L, Araujo MAM, Amaral A et al. (2005) Cell signaling pathways in Paracoccidioides brasiliensis – inferred from comparism with other fungi. Genet Mol Res 4:216–231
Galagan JE, Selker EU (2004) RIP: the evolutionary cost of genome defense. Trends Genet 20:417–423
Gessler NN, Aver'yanov AA, Belozerskaya TA (2007) Reactive oxygen species in regulation of fungal development. Biochemistry (Mosc) 72:1091–1109
Glass NL, Jacobson DJ, Shiu PKT (2000) The genetics of hyphal fusion and vege-tative incompatibility in filamentous ascomycete fungi. Ann Rev Genet 34:165–186
Glass NL, Kaneko I (2003) Fatal attraction: nonself recognition and heterokaryon incompatibility in filamentous fungi. Eukaryot Cell 2:1–8
Glass NL, Saupe SJ (2002) Vegetative incompatibility in filamentous ascomycetes. In: Osiewacz HD (ed) Molecular Biology of Fungal Development. Marcel Dekker, New York
Griffin DH (1994) Fungal Physiology. 2nd ed. Wiley-Liss, New York
Gross L (2006) A chemical facilitator of plant-fungus communication. PLoS Biology 4:e238–e242
Hamann A, Brust D, Osiewacz HD (2008) Apoptosis pathways in fungal growth, development and aging. Trends Microbiol doi:10.1016/j.tim.2008.03.003

Hartmann HA, Kahmann R, Bölker M (1996) The pheromone response factor coordinates filamentous growth and pathogenicity in ustilago maydis. EMBO J 15:1632–1641

Hartmann HA, Krüger J, Lottspeich F et al. (1999) Environmental signals controlling sexual development of the corn Smut fungus Ustilago maydis through the tran-scriptional regulator Prf1. Plant Cell 11:1293–1305

Hemenway CS, Heitman J (1999) Calcineurin: structure, function, and inhibition. Cell Biochem Biophys 30:115–151

Herranz S, Rodriguez JM, Bussink HJ et al. (2005) Arrestin-related proteins mediate pH signalling in fungi. Proc Natl Acad Sci U S A 102:12141–12146

Hoffman CS (2005) Except in every detail: comparing and contrasting G-Protein signaling in Saccharomyces cerevisiae ans Schizosaccharomyces pombe. Eukaryot Cell 4:495–503

Hogan DA (2006) Talking to themselves: autoregulation and quorum sensing in fungi. Eukaryot Cell 584:613–619

Hohmann S (2002) Osmotic stress signalling and osmoadaptation in yeasts. Microbiol Mol Biol Rev 66:300–372

Humber RA (2008) Evolution of entomopathogenicity in fungi. J Invertebr Pathol 98:262–266

Jakupovic M, Heintz M, Reichmann P et al. (2006) Microarray analysis of expressed sequence tags from haustoria of the rust fungus Uromyces fabae. Fungal Genet Biol 43:8–19

Kays AM, Borkovich KA (2004) Signal transduction pathways mediated bxy heterotrimeric G proteins. In: Brambl R, Marzluf GA (eds) The Mycota III. Springer, Berlin

Kellis M, Birren BW, Lander ES (2004) Proof and evolutionary analysis of ancient genome duplication in the yeast Saccharomyces cerevisiae. Nature 428:617–624

Kellis M, Patterson N, Endrizzi M et al. (2003) Sequencing and comparism of yeast species to identify genes and regulatory elements. Nature 423:241–254

Kopper BJ, Klepzig KD, Raffa KF (2004) Components of antagonism and mutualism in Ips pini – fungal interactions: relationship to a life history of colonizing highly stressed and dead trees. Environ Entomol 33:28–34

Krantz M, Becit E, Hohmann S (2006) Comparative genomics of the HOG-signalling system in fungi. Curr Genet 49:137–151

Krüger J, Loubradou G, Regenfelder E et al. (1998) Crosstalk be-tween cAMP and pheromone signalling pathways in Ustilago Maydis. Mol Gen Genet 260:193–198

Lammers PJ (2004) Symbiotic signaling: new functions for familiar proteins. New Phytol 161: 324–326

Lang BF, O'Kelly C, Nerad T et al. (2002) The closest unicellular relatives of animals. Curr Biol 12: 1773–1778

Lee N, D'Souza CA, Kronstad JW (2003) Of smuts, blasts, mildews, and blights: cAMP signaling in phytopathogenic fungi. Ann Rev Phytopathol 41:399–427

Lengeler KB, Davidson RC, D'Souza C et al. (2000) Signal Transduction cascades regulating fungal development and virulence. Microbiol Mol Biol Rev 64:746–785

Levin D (2005) Cell wall integrity signalling in Saccharomyces cerevisiae. Microbiol Mol Biol Rev 69:262–291

Levina NN, Lew RR (2006) The role of tipp-localized mitochondria in hyphal growth. Fungal Genet Biol 43:65–134

Lott TJ, Fundyga RE, Kuykendall RJ et al. (2006) The human commensal yeast, Candida albicans, has an ancient origin. Fungal Genet Biol 42:444–451

Margulis L, Schwartz KV (1988) Five Kingdoms. W. H. Freeman and Company, New York

McAlester G, O'Gara F, Morrissey JP (2008) Signal-mediated interactions between Pseudomonas aerugiunosa and Candida albicans. J Med Microbiol 57:563–569

Mitchell TK, Dean RA (1995) The cAMP-dependent protein kinase catalytic subunit is required for appressorium formation and pathogenesis by the rice blast pathogen Magnaporthe grisea. Plant Cell 7:1869–1878

Moradas-Ferreira P, Costa V (2000) Adaptive response of the yeast Saccharomyces cerevisiae to reactive oxygen species: defenses, damage and death. Redox Report 2000 5:277–285

Muirhead CA, Glass NL, Slatkin M (2002) Multilocus self-recognition systems in fungi as a cause of trans-species polymorphism. Genetics 161:633–641

National Institute of Allergy and Infectious Diseases (NIAID) (1993) Molecular Medical Mycology SECRETION. Proceedings of the NIAID Workshop in Medical Mycology. University of Minnesota, Minneapolis, June 24–26, 1993 http://www.niaid.nih.gov/dmid/meetings/mycology/secretion.htm

Poulsen M, Boomsma JJ (2005) Mutualistic fungi control crop diversity in fungus-growing ants. Science 307:741–744

Prusty R, Grisafi P, Fink GR (2004) The plant hormone indoleacetic acid induces inva-sive growth in Saccharomyces cerevisiae. Proc Natl Acad Sci U S A 101:4153–4157

Raven JA (2001) Selection pressures on stomatal evolution. New Phytol 153:371–386

Reynolds TB, Fink GR (2001) Bakers' yeast, a model for fungal biofilm formation. Science 291:806–807

Rooney AP, Ward TJ (2005) Evolution of a large ribosomal RNA multigene family in filamentous fungi: birth and dead of a concerted evolution paradigm. Proc Natl Acad Sci U S A 102:5084–5089

Sanders WB (2001) Lichens: interface between mycology and plant morphology. Bioscience 51:1025–1035

Sanders WB (2006) A feeling for the superorganism: expression of plant form in the lichen thallus. Bot J Linn Soc 150:89–99

Santos JL, Shiozaki K (2004) Phosphorelay signaling in yeast in response to changes in osmolarity. Sci Signal 262 DOI: 10.1126/stke.2622004tr12

Schultze K, Schimek C, Wostemeyer J et al. (2005) Sexuality and parasitism share common regulatory pathways in the fungus Parasitella parasitica. Gene 348:33–44

Schwarze FWMR, Engels J, Mattheck C (2004) Fungal Strategies of Wood Decay in Trees. 2nd ed. Springer, Heidelberg

Selker EU, Tountas NA, Cross SH et al. (2003) The me-thylated component of the Neurospora crassa genome. Nature 422:893–897

Sullivan BT, Berisford CW (2004) Semio-chemicals from fungal associates of bark beetles may mediate host location behavior of parasitoids. J Chem Ecol 30:703–717

Tan Y, Merrow M, Roenneberg T (2004) Photoperiodism in Neurospora Crassa. J Biol Rhyth 19:135–143

Van Diepeningen AD, Debets AJM, Hoekstra RF (2006) Dynamics of dsRNA mycoviruses in black aspergillus populations. Fungal Genet Biol doi:10.1016/j.fgb.2006.01.014

Villarreal LP (2005) Viruses and the Evo-lution of Life. ASM Press, Washington

Villarreal LP (2009) Origin of Group Identity. Viruses, Addiction and Cooperation. Springer, New York

Volk T (2002) The humongous fungus–ten years later. Inoculum 53:4–8

Wang P, Heitman J (1999) Signal transduction cascades regulating mating, filamentation and virulence in Cryptococcus neoformans. Curr Opin Microbiol 2:358–362

Weld RJ, Plummer KM, Carpenter MA et al. (2006) Approaches to functional genomics in filamentous fungi. Cell Res 16:31–44

Whisson SC, Avrova AO, Lavrova O et al. (2005) Families of short interspersed elements in the genome of the oomycete plant pathogen, Phytophtora infestans. Fungal Genet Biol 42:351–365

Witzany G (2006) Plant Communication from biosemiotic perspective. Plant Sign Behav 1: 169–178

Witzany G (2007) The *Logos* of the *Bios* 2. Biocommunication. Umweb, Helsinki

Wu J, Glass NL (2001) Identification of specificity determinants and generation of alleles with novel specificity at the het-c heterokaryon incompatibility locus of Neurospora crassa. Mol Cell Biol 21:1045–1057

Xue C, Hsueh YP, Heitman J (2008) Magnificent seven: roles of G protein-coupled receptors in extracellular sensing in fungi. FEMS Microbiol Rev 32:1010–1032
Ingold CT, Hudson HJ (1993) The Biology of Fungi. Chapman & Hall, London
Jennings DH (1995) Physiology of Fungal Nutrition. Cambridge University Press, New York
Yu RC, Pesce CG, Colman-Lerner A et al. (2008) Negative feedback that improves information transmission in yeast signalling. Nature 456:755–761

Chapter 6
Bacteria Communication

Abstract Communicative competences enable bacteria to develop, organise and coordinate rich social life with a great variety of behavioural patterns. Biocommunication of bacteria is the presupposition even for behavioural patterns in which they organise themselves like multicellular organisms. They have existed for almost four billion years and still survive, being part of the most dramatic changes in evolutionary history such as DNA invention, cellular life, invention of nearly all protein types, partial constitution of eukaryotic cells, vertical colonisation of all eukaryotes, high adaptability through horizontal gene transfer and co-operative multispecies colonisation of all ecological niches. Recent research demonstrates that these bacterial competences derive from the aptitude of viruses for natural genome editing. Bacteria seem to be the optimal biotic matrix for virus-induced genetic inventions.

6.1 Introduction

In this chapter I describe recent findings on bacteria, how they interpret and coordinate, what semiochemical vocabulary they share and which goals they try to reach. In a second stage I describe the main categories of (sign-mediated) interactions between bacterial and non-bacterial organisms, and between bacteria of the same or related species, and natural genetic engineering within bacteria. In a third stage I will focus on the relationship between bacteria and their obligate settlers, i.e. viruses. We will see that bacteria are important hosts for multiviral colonisation and virally-determined order of nucleic acid sequences.

6.1.1 Biocommunicative Competences of Bacteria

Bacteria (prokaryotes) communicate and therefore are able to organize and coordinate their behaviour similar to a multicellular organism (Ben Jacob and Levine 2006, Bassler and Losick 2006). Communication processes are sign-mediated

G. Witzany, *Biocommunication and Natural Genome Editing*,
DOI 10.1007/978-90-481-3319-2_6, © Springer Science+Business Media B.V. 2010

interactions. Signs are, in most cases, chemical molecules (in some cases also tactile interactions) which serve as signals both within and between prokaryotic organisms. Bacteria are symbiotic organisms covering the whole range from mutualism to parasitism. They may be beneficial for their (eukaryotic) hosts and without them host survival would not function. Others are neutral, i.e. they do not harm the host. Many of them also cause diseases, with sometimes epidemic characteristics and, often, lethal consequences.

Bacteria represent one of the main success stories of evolution. They originated at the early beginning of life similarly to archaea which represent a different organismic kingdom (Woese et al. 1990). Bacteria are found in all ecological niches, and share a common flux of their gene pool with a high rate of gene order recombination for adaptational purposes of great diversity (Pal et al. 2005). More than in any other organismic kingdom it is in common use to speak about the languages of bacteria (Kaiser and Losick 1993; Swift et al. 1994; Bassler 1999, 2002; Schauder et al. 2001; Schauder and Bassler 2001; Ben Jacob et al. 2004).

Quorum sensing is the term of description for sign-mediated interactions in which chemical molecules are produced and secreted by bacteria (Crespi 2001, Manefield and Turner 2002, Greenberg 2003). They are recognized by the bacterial community dependent on a critical concentration and in a special ratio to the population density (Daniels et al. 2004; Waters and Bassler 2005). These molecules trigger the expression of a great variety of gene transcriptions. Many bacteria use multiple quorum sensing codes; each may be modulated by post-transcriptional or other regulatory engineering (Loh et al. 2002).

There are also communication processes between different species of bacteria and between bacteria and non-bacterial life such as eukaryotic hosts (Konaklieva and Plotkin 2006). Beneath the semiochemicals necessary for developmental processes of bacterial communities such as division, sporulation and synthesis of secondary metabolites, there are physical contact-mediated behavioural patterns which are important in biofilm organization (Davis et al. 1998; Fuqua and Greenberg 2002; Voloshin and Kaprelyants 2004; Parsek and Greenberg 2005). Also, abiotic influences serve as signs which indicate specific nutrients or other environmental circumstances such as hydrodynamic changes.

As communities of bacteria species, which are able to coordinate their behaviour and have advantages over single bacteria organisms, are much more common, it is not surprising that the evolutionary drive went into rising communicative complexity (Ben Jacob 2003). We should not forget that in comparison to the first two billion years of life on earth with closed prokaryotic symbiology the rise and growth of the multicellular eukaryotes (animals, fungi, plants) was a *crucial advantage* for bacterial lifestyle to colonize vertical hosts with their great spatial and motility resources.

We can differentiate three classes of signalling molecules for different purposes, i.e. signalling within the organism to coordinate gene expressions to generate adequate response behaviour, signalling between the same or related and different species. With a limited number of molecules and a limited number of combinatorial rules they generate quite different interactions for different purposes all mediated by signs. As in every sign-mediated interaction sign users share a common set of syntactic rules, i.e. how signs may be combined; of pragmatic rules which determine a great variety of interactional contexts, e.g. development, growth, mating, virulence, attack and defence. The situational context of these complex interactional processes determines the meaning of signs, i.e. semantics of signals. Independent of organismic complexity the complementarity of these three levels of semiotic rules can be identified, in principle, in every sign-mediated interaction within and between organisms (Witzany 2006b, 2007). This leads to the generation of intra- and intercellular processes which enable bacterial communities to generate memory which may be inheritable but can alter epigenetically, i.e. different reading/meaning patterns of the same genetic data set with differences at the phenotypic level without altering the genetic data set.

The link between linguistics and genetics has been obvious since the detection of the universal grammar and the structural code of DNA. Chomsky's meaning-independent syntax approach lead to the broad acceptance and usage of bioinformatic methods and systems biology. Researchers in bacteria communication like Ben Jacob et al. (2004) suggested with good reason that this approach reduces linguistic competences found in bacterial communication and has to be satisfied by both semantic aspects, i.e. the context-dependent meaning of signals which act as signs, and pragmatic aspects, which focus on the variety and differences of the behavioural patterns in common-goal coordination, shared knowledge, memory and mutual intentions. Apart from that, it is coherent with the presupposition by Charles Morris of any non-reductionistic analysis of language-like structures, the complementarity of syntax, semantics and pragmatics.

6.1.2 Biofilm Organisation: Interpretation and Coordination

Bacteria have profound effects on human health, agriculture, industry and other ecospheres. Therefore they target the multiple drugs which fight them (Camara et al. 2002). They develop drug resistance by coordination of special defensive behavior called biofilm organization. Biofilm organization is a special kind of coordination with a high density of physical contact and contact-specific signalling (Bassler and Losick 2006). If bacteria realize a critical mass via quorum sensing they organize a high density of communal body by moving their flagellas which may resist even strong antibiotics (Wadhams and Armitage 2004). Biofilms are constructed on abiotic surfaces, e.g. on stones in rivers and other aqueous surfaces, as well as biotic ones, e.g. in the respiratory track of animals. Each human who had a strong cough remembers like persistent the mucus in the bronchial tube remained.

Nutrient availability also regulates the structure of biofilm organization (Stanley and Lazazzera 2004) as hydrodynamic forces (Wuertz et al. 2004). Interestingly, it has been found that biofilm organization is linked with coordinated DNA release which is integrated in the biofilm (Spoering and Gilmore 2006).

6.2 Semiochemical Vocabulary and Communicative Goals

The semiochemical vocabulary used by bacteria is of great variety, especially because some signalling molecules are multiple reusable components (Henke and Bassler 2004). Acyl homoserine lactones (AHLs) and linear oligopeptides are used as signs in diverse processes. Cyclized oligopeptides function as virulence genes. g-Butryolactones (GBLs) are used as antibiotics and in sporulation processes. Furanosyl diester (AI-2) is used in diverse processes (Sun et al. 2004) and in luminescence. cis-11-Methyl-2-dodecenoic acid (DSF) serves in virulence and pigmentation. 4-Hydroxy-2-alkyl quinolines (PQS, HAQs) are important in whole regulation processes and for virulence as are palmic acid methyl esters (PAME). Putrescine is important in swarming motility like biofilm organization. A-signal is used in early developmental processes and aggregation. C-signal is a cell surface-associated protein and serves to coordinate motility and the developmental process of building a fruiting body. Cyclic dipeptide is a secondary metabolite (Shapiro 1998; Visick and Fuqua 2005) (Fig. 6.1).

Gram-negative bacteria use homoserine lactones (LuxR/LuxI) as signs in communication processes (Swift et al. 1994; Lenz et al. 2004), whereas Gram-positive bacteria use oligopeptides in quorum sensing communication. As in all organisms non-coding RNAs are important in higher order regulatory pathways. Small RNAs and microRNAs are used by bacteria to regulate special genetic expression patterns which play an important role as appropriate response behavior to stress or nutrient

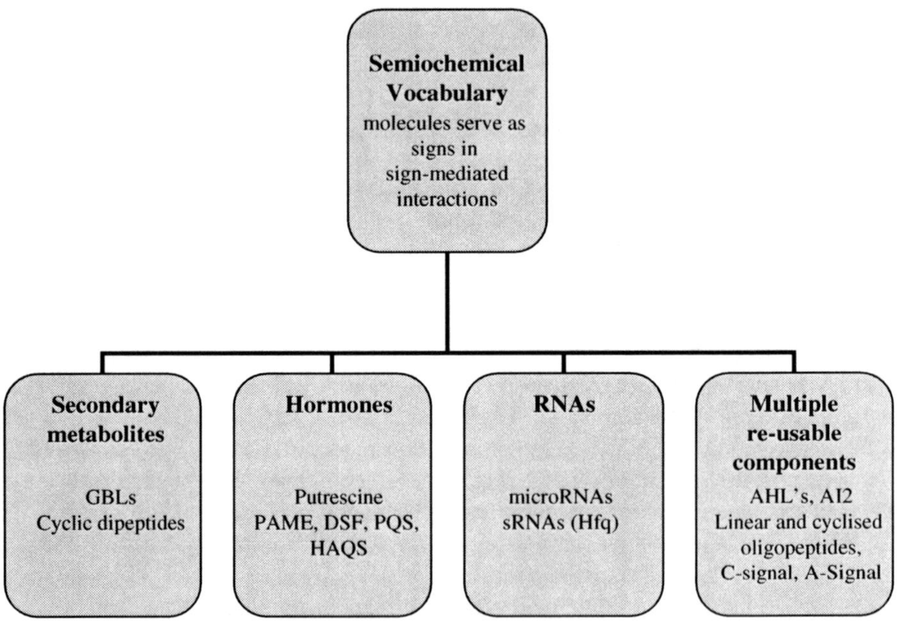

Fig. 6.1 Some semiochemicals used in bacterial communication processes

availability (Teplitski et al. 2000; Vogel and Sharma 2005), e.g. in controlling the quorum sensing pathways (Bauer and Robinson 2002).

At present, three kinds of communicative goals are distinguished: (A) reciprocal communication, active sign-mediated interactions which are beneficial for both interacting parts; (B) messages which are produced as response on a triggering event which may be an indicator for a receiver which was not specially targeted by the producer. A coincidental event which is neutral – except for the energy costs of production – to the producer but beneficial for the receiver; (C) signalling to manipulate the receiver, i.e. to cause a response behavior which is one-sided – beneficial to the producer and harmful to the receiver (Visick and Fuqua 2005), often in that they behave against their normal goals (Keller and Surette 2006).

The three classes of intra-, inter- and transorganismic (trans-specific) communication enable bacteria to generate and coordinate different behavioral patterns: self and non-self identification, i.e. 'recognition' and identification of other colonies and measurement of their size, pheromone-based courtship for mating, alteration of colony structure in formatting of fruiting bodies, initiation of developmental and growth processes, e.g. sporulation.

In receiving signals from same or related species or non-bacterial organisms the signalling molecules bind to specialized sensor proteins which function as receptors. They transmit the message to an intracellular regulator (Fuqua et al. 1996; Visick and Fuqua 2005), i.e. the signal molecule transits the cell membrane through diffusion or by specific transport pathways. Inside the cell the signalling molecule, in most cases, binds to a cytoplasmic target protein. It may be that a diffusible molecule is chemically engineered to an active signal after entering the target cell (Visick and Fuqua 2005). Organization of cellular production of response molecules leads to signal-dependent transcription control of DNA.

Bacteria have to distinguish between species-specific signalling and signalling which is able to modulate behaviors interspecifically (Bassler 1999). With these communicative competences they are able to coordinate species-specific behavourial patterns as well as to coordinate behaviors between diverse species.

6.3 Transorganismic Communication

Starting with beneficial symbioses between bacteria and plants we refer to the complex communication networks between soil bacteria, mycorrhizal fungi and plant roots (Hayashi 2005). Mycorrhizal fungi secret molecules in the surrounding environment which serve as nutrients for soil bacteria and trigger their activation to degrade special nutrients which are then available for mycorrhizal fungi. Their hyphal growth serves as the developmental and growth area of plant roots, themselves being dependent on nutrients which are prepared by the mycorrhizal fungi. Plant roots can also mimic bacterial signalling molecules, either to trigger bacterial production of special molecules or to disturb bacterial communication pathways (Teplitski et al. 2000; Bauer and Robinson 2002; Daniels et al. 2004).

Rhizobia bacteria are integrated into plant cells by phagocytosis when they interact symbiotically with the plant roots (Samaj et al. 2004). In other cases where rhizobia fail to fix nitrogen inside the root nodules because they are being deceptive, plants are sanctioning these rhizobia (Kiers et al. 2003) and prevent their spread in order to stabilize mutualistic symbioses with bacterial colonies (Keller and Surette 2006). Root exudates of different kinds regulate plant and microbial communities in the rhizosphere. This is necessary to stabilize equilibrium and inhibit the continuity of attacks by pathogenic bacteria in the soil (Walker et al. 2003; Bais et al. 2003). The full range of trans-specific communication processes between bacteria and plant roots are important for developmental and growth processes in the entire plant kingdom (Manefield and Turner 2002; Sharma et al. 2003).

Chemical molecules which serve as signs in intercellular communication processes of bacteria are similar to pheromones in social insects and animals. This may be an indicator of evolutionary lineages that evolved in the bacterial 'chatter' (Velicer 2003) (Fig. 6.2).

Marine eukaryotes are able to mimic bacterial quorum sensing to inhibit bacterial successful communication (Rice et al. 1999). Interbacterial communication uses hormone-like signalling to sense specific host locations such as intestinal habitat. In this specialized ecosphere a bacteria–host communication occurs which means the host cells and bacterial cells share a common meaning function for the same signalling molecules (Sperandio et al. 2003).

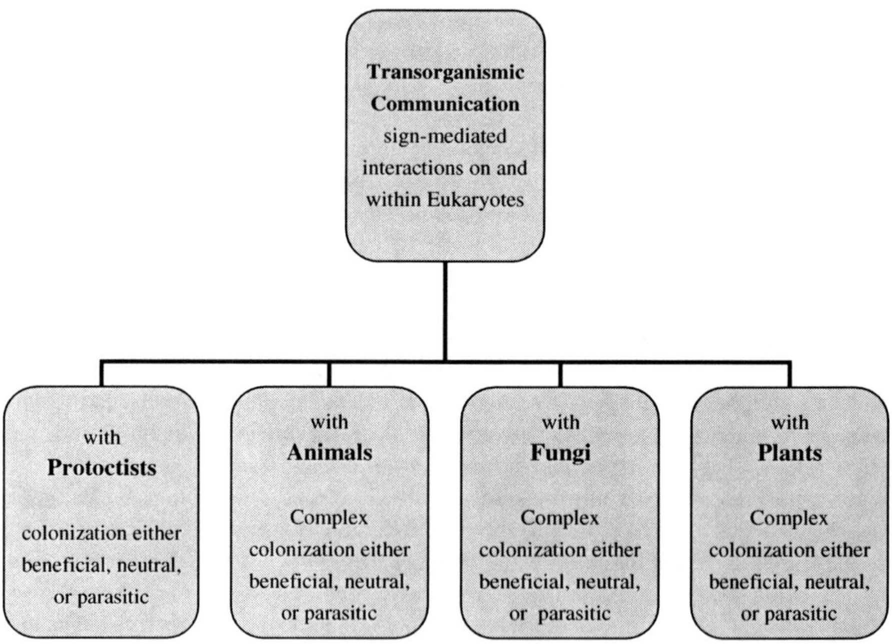

Fig. 6.2 Transorganismic (trans-specific) communication levels in bacterial communication

Living as endosymbionts as potential candidates for symbiogenesis (Margulis 1996, 1999, 2004, Margulis and Sagan 2002), as documented in the origin of eukaryotic endosomes like mitochondria, indicates the important role of bacteria for the entire history of evolution (Witzany 2005). The interactions may be pericellular colonization events but also an intracellular lifestyle. These different symbiotic interactions range from acquisition of novel genetic material to reduction in size and content connected with gene loss (Batut et al. 2004). Successful living processes of higher eukaryotes would not be viable without beneficial symbiosis with bacteria. The cell mass of an adult human assembles 20% of human origin and up to 80% of exogenic settlers (Blech 2000), most of them bacteria.

6.4 Interorganismic Communication

For a long time it was assumed that bacteria live predominantly as monads. However, it has been recognized that this is a very rare exception (Federle and Bassler 2003; Dunn and Handelsman 2002). Bacterial colonies live, in almost all cases, not alone but in coexistence with other bacterial species self-coordinated by a diversity of sign-mediated interactions (Gray 1997; McNab and Lamont 2003; Ben Jacob and Levine 2006). Bacteria use intraspecific and interspecific signalling in all ecological in vivo situations (Keller and Surette, 2006). This also implies a broad variety of conflicts within and between species (Velicer et al. 2000). The mutual, neutral and manipulative aims of communication processes are special kinds of response behavior to certain degrees of beneficial up to conflictual relationships (Keller and Surette 2006).

Dependent on the availability of nutrients, some bacteria suppress normal cell development which leads to the development of a different cell type which is better suited for adequate response behavior for this situational context. It means that different environmental conditions can lead to different gene expressions within the same genomic data set. It has been shown that if the same colony is exposed several times to these changing contexts they react more immediately. This indicates that bacterial communities are able to develop collective memory and learn from the experience (Ben Jacob et al. 2004; McNab and Lamont, 2003). This functions similar to neuronal networks in higher eukaryotes. In the case of changing environmental conditions, the suppression of cell division may lead to cell elongation which enables cell colonies to change the modus of motility. This is an important feature of socio-bacterial behavior, e.g. swarming coordination and organization for surface colonization (Shapiro 1998, 2007) (Fig. 6.3).

Some authors have documented altruistic strategies in mixed colony formations which seems to be an advantage to the mixing among microcolonies. Altruistic behavioral strategies enable strengthened self-identity and a sustainable equilibrium in multilevel colonized ecological niches (Velicer and Yu 2003; Kreft 2004).

Interestingly, bacteria use a common contextual interpretation of incoming signals by each member of the colony. The response behavior is appropriate to the majority vote (Ben Jacob et al. 2004) of the context-dependent decision.

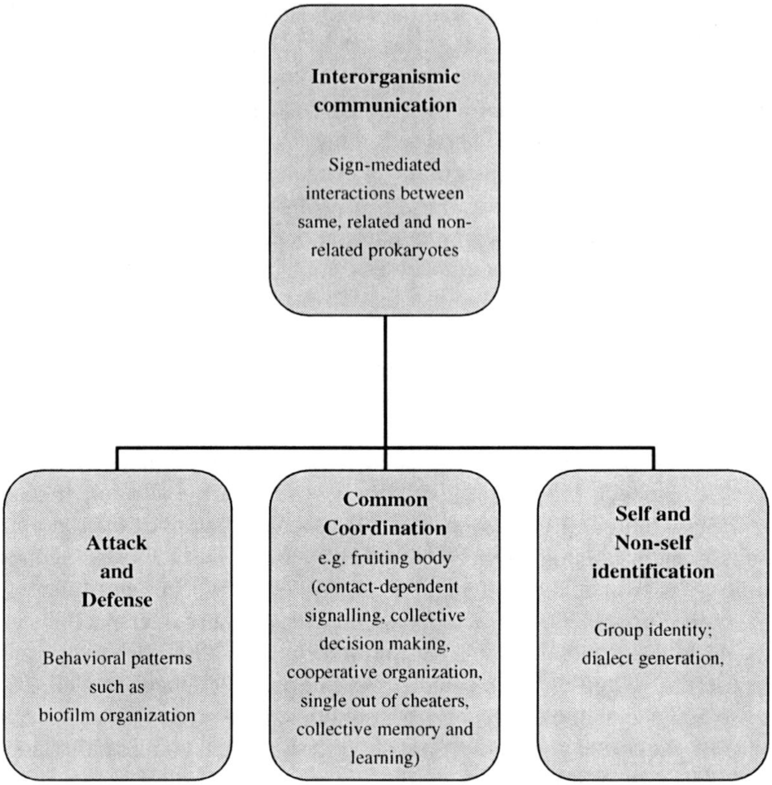

Fig. 6.3 Species-specific, species-related and non related prokaryotic communication

The identification of non-self species is a competence which is possible through species-specific and group-specific quorum sensing and is coherent with the assumption that smaller groups of the same bacterial species are able to built types of quorum-sensing 'dialects'. These are important in the high density of coexistent bacterial life habitats to prevent confusion and enable more complex coordination (Taga and Bassler 2003). Interestingly, the prokaryotic cell–cell communication has structural analogues to cross-kingdom signalling between bacteria and fungi (Wang et al. 2004).

Bacteria decide, in special cases, to form fruiting bodies of different types and shapes for sporulation. This enables bacterial communities to more efficiently disseminate the spores. The fruiting body building is governed by context-specific rules with different roles for different sub-groups of bacterial communities for coordination (Kaiser and Welch 2004). Some have to serve for motility to density, followed by direction decision and decision of cell types, cell growth and developmental stages in all the different steps until the fruiting body is ready for the sporulation event. Without communicative hierarchical organization this would not be possible. If communication is disturbed body building is not assured, so bacterial communities have developed special strategies to single out so-called 'cheaters'

(Velicer et al. 2000; Ben Jacob et al. 2004), which do not follow the rules for coordinating this special behavior.

One of the most interesting and best investigated phenomena of bacterial communication is the *symbiology* of multiple colonies coexistent in the human oral cavity (Kohlenbrander et al. 2002, 2005; Rickard et al. 2006). Bacteria on human teeth and oral mucosa establish a homeostasis of pathogenic and mutualistic bacteria by a complex system of sign-mediated interactions both species-specific and trans-specific. The dental plaque in the oral cavity of humans is a unique habitat which is not found in any other species (Sahasrabudhe and Dewhirst 2001). The homeostasis is not static but is the result of a dynamic relationship between different species-colonies dependent on intervals of daily hygiene. The interacting species number approximately 500 different species (Moore and Moore 1994; Kroes et al. 1999; Paster et al. 2001).

Each member of these communities must be capable of self and non-self distinction, and be able to distinguish between species-specific signalling and trans-specific signalling or even 'noise' (no biotic content). As a community they must be able to measure their own colony size and the size of the other colonies and molecules that have the same chemical structure but are not part of a biotic message. Special communication patterns with detailed hierarchical steps of signal production and transmission include (i) metabolite exchange, (ii) cell–cell recognition, (iii) genetic exchange, (iv) host signal recognition and signal recognition of same or related species. Owing to the high number of competing and cooperating species there is a special short- and long-term community architecture established. If the communication on the intra-, inter- and transorganismic level is successful, i.e. the signal transmission and reception enables colonies to live in a dynamic homeostasis, then the human oral cavity will avoid cavity diseases (Kohlenbrander et al. 2002, 2005).

6.5 Intraorganismic Communication

Interestingly, prokaryotic gene order is not as conserved as the protein sequences. Only some higher order regulations (operons) that code for physically interacting proteins are found in almost all bacterial (and archaeal) genomes. Recent research indicates high dynamics of new gene orders as documented in the horizontal gene transfer events with their intensive intragenomic recombination (Immaizumi-Anraku et al. 2005; Xie et al. 2004). This exchange of whole genes or gene-blocks enables bacterial lifestyles to combine several bacterial competences, i.e. phenotypes. The transformation process includes the release of naked DNA, followed by the uptake and recombination, i.e. the integration, with 17 steps identified to date (Thomas and Nielsen 2005). Thus we can recognize the outcomes of a diversity of mobile DNA contents (Bordenstein and Reznikoff 2005), not a mass of individualized genetic texts, but a bacterial gene pool as a text repertoire which is available for each individual bacteria and the resource for bacterial genome innovation and evolution (Gogarten and Townsend 2005). Horizontal gene transfer is a main resource for integrating newly evolved genes into existing genomes and does not need the slow steps of chance mutations to alter the genomes but accelerated genome innovations in both bacteria and archaea (Jain et al. 1999, 2003; Brown 2003). Important in this context of genomic innovation is not the sequence acquisition alone but also the contextualization (Solomon and Grossman 1996); it means also their loss (Berg and Kurland 2002). It seems now that the phylogeny of microbial species is not a tree of life, but an evolutionary network or a ring of life, mediated by genetic exchange, i.e. acquisition and loss of genetic data sets (Rivera and Lake 2004; Kunin et al. 2005).

6.5.1 Intracellular Communication

Signal-dependent transcription regulation of the DNA serves for a great variety of response behavior. One of the most interesting phenomena is the fact that in the first two billion years of life on planet earth the immense density of bacterial life has not been an event of the mass of individual organisms but their commonly shared gene pool which was in constant flux, as we now know, through investigations on horizontal gene transfer. It means that the evolution of bacteria was not a random event of chance mutations and their selection but transfer of whole genes and gene-blocks representing real phenotypes that were transferred. This leads to different combinatorial patterns of genetic encoded phenotypes and the rise of bacterial diversity. It also enables bacterial pathogens to optimize their disease-causing coordination and is therefore targeted to special kinds of drug developments for medical purposes (Tettelin et al. 2005). New empirical data seem to suggest that the phenomenon of horizontal gene transfer is driven by viral competences inherent in bacterial settlers such as phages, plasmids, retroplasmids and transposons (Villarreal 2005) (Fig. 6.4).

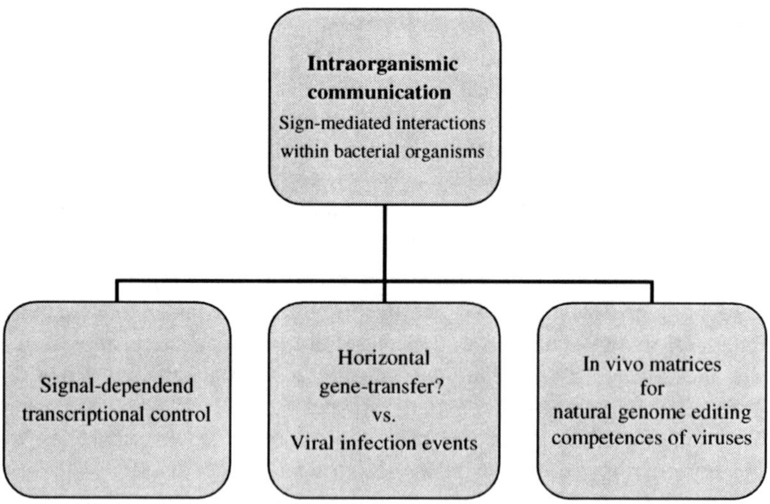

Fig. 6.4 Communication processes within prokaryotes

For a long time it has been proposed that tubulin plays an important role in cytoskeletal functions of eukaryotes whereas prokaryotes lack this system. Recent research has shown that tubulin is a very ancient system for genetic data set segregation also in bacteria which plays important roles in filament formation, movement and orientation (Graumann 2004; Graumann and Defeu-Soufo 2004; Defeu-Soufo and Graumann 2004; Gitai 2005; Guerrero and Berlanga 2007).

6.5.2 Bacterial Evolution and the Agents of Natural Genome Editing

To elucidate communicative competences of bacteria we also have to look at the roles of viruses and their relationship to bacteria. Viruses have long been accepted only as disease causing, epidemic phenomena with lytic and therefore extremely dangerous consequences for infected organisms. However, new research has corrected this picture. Viruses are part of the living world, in most cases integrated in the cytoplasm or the nucleoplasma of cells without harming the host. Viruses are on their way to representing the best examples of symbiotic relationships, because there is no living being since the start of life that has not been colonised by them, in most often cases in the form of multiple colonisations (Witzany 2006a). The longest period of these symbiotic relationships during evolutionary history share viruses, archaea and bacteria. As viruses are extremely biosphere specific, i.e. they adapt to special host tissues, the identification of various forms of, e.g. bacteria is to identify primarily the viruses that colonise them. This is also the concept of 'bacteriophages', in that bacteria are identified best by identifying the viruses that are associated with them. Host identification in this way is a special method called phage typing.

6.5.3 Lytic versus Persistent Viral Life-Strategies

As mentioned in recent years and outlined in the next chapter in greater details the lytic consequences of viral infection are a special case if viruses are not able to develop a sessile lifestyle without harming the host. In most cases viruses living within organisms help to ward off competing parasites from the host and becoming part of its evolutionary history. Persistent viruses are decisive for species diversity and host genome editing. Nearly all cellular key processes such as expression, transcription, translation and recombination with all their detailed steps derived from viral competences. Even the DNA replication pathways, after a period of early RNA influence (Forterre, 2002, 2005, 2006), seems to be a special viral strategy for the conservation of coded phenotypes by warding off RNA parasites (Villarreal and DeFillipis, 2000; Villarreal 2005).

Since observations have become more evident that viruses are able to integrate genetic material into the host genome, it has become clear that viruses have infection lifestyles and endosymbiotic and even symbiogenetic lifestyles. They bestow phenotypic capabilities on the host which non-infected hosts from the same species do not possess. As endosymbiotic viruses which are dependent on the host's replication they are part of the host history in that they are inheritable and part of the genomic identity of the host as documented in some several 10,000 infection events in the human genome by endogenous retroviruses (Villarreal 2004).

The two viral lifestyles are not in strict opposition but, in most cases, are part of a symbiotic process. It starts with an infection by a virus. In the infected host it arrives at an equilibrial status where the immune system does not eliminate the virus but controls its replication without fatal consequences for the host organism. The persistent status lasts during most phases of the host's life, but may return to the lytic lifestyle if the host-immune system is under stress (Villarreal et al. 2000). Most often the integration occurs by mutual neutralisation of toxic capabilities by an antitoxin of a competing genetic settler (Pandey and Gerdes 2005). The whole range of toxin/antitoxin addiction modules we can find througout all genetic contents in living nature most likely is of viral origin (Villarreal 2005). Therefore the persistence is sometimes called temperate lifestyle. A good example is the persistent virus in all Symbiodinium species being the essential endosymbiotic partner for coral animals. Coral bleaching as a worldwide phenomenon of coral disease is the consequence of dying of the coral endosymbiont because of global (water) warming. As we know now, death occurs because the persistent viruses of Symbiodinium become lytic as a reaction to the changing water temperature (Villarreal 2005).

Also bacteria may be infected by viruses without being harmed. If infected bacteria meet non-infected bacteria it may be that the non-infected acquires lysis; the lysogenic strain does not lyse itself, but is lethal to the non-infected one. The colonized bacteria has a virus-derived molecular genetic identity which has an advantage against the non-infected one through an acquired ability. This lysogenic bacteria, termed prophage, has an immunity function for the bacteria which the non-infected bacteria lack. Prophage is a virus that is integrated into the bacterial host genome. Both the acute lytic phages and the persistent prophages such as T4 and lamda are highly abundant in oceans and in the soil and seem to be the most dynamic life form on the entire planet. Some viruses are not integrated in the host genome but persist in the cytoplasma and replicate independently from the host genome (Villarreal 2005).

When we speak about the relationship of bacteria and viruses in most cases we speak about phage ecology. Most prokaryotic viruses are double-stranded DNA viruses with either linear or circular genome morphology and are packaged in an icosahedral capsid. Whereas acute viruses in most cases code for their own replication, recombination and repair proteins, the persistent phages lack such genes and use the host-cellular replication. This involves a totally different gene word order (Villarreal 2005) in acute lytic and in persistent phages. This is documented in the very different nucleotide words (di-, tri- and tetranucleotides). Nucleotide word frequences of acute phages are very dissimilar to those of their hosts while persistent or temperate phages share nucleotide word frequences with the host. This means the molecular syntax from acute and persistent phages is constructed totally differently according to the different strategies. Different life strategies with different behavioral pathways need a completely different semantic content in the genome expressed in a different syntactic arrangement of nucleotides (Witzany, 2006b).

As the bacterial cell walls differ substantially between different types of bacteria a different behavior is necessary for viruses for recognition, attachment and penetration. Owing to these diverse barriers of the bacterial cell walls, the prokaryotic viruses do not enter the host cells physically but attach to the cell surface and inject their genomes through contractile tails or pilot proteins. Also, the progeny of the virus has to deal with this barrier (Villarreal, 2005).

Bacterial DNA does not have highly stable structures as do eukaryotes and can interact with the cellular replication and transcription. In most cases it is circular with a unique origin of replication system. In contrast to that viral double-stranded DNA is a linear DNA with integrated short terminal repeats. Since bacterial viruses do not use a transport technique as they need in eukaryotes to be transported out of the nucleus, bacterial viruses differ a great deal from eukaryotic viruses.

All bacteria have a restriction/modification system which is a connected form of two viral competences. Only the descendants of mitochondria lack this system which causes them not to be exposed to viral selection. It may be that they have transposed their ability to the eukaryotic nucleus which cares in a more efficient way for cell immunity (Villarreal 2005)

6.5.4 Bacteria as Biotic Matrix for Natural Genome Editing

Horizontal gene transfer between bacteria as being responsible for genetic plasticity in prokaryotes may be a capability which is acquired by viral infections. Then, viral genetic inventions are transferred not from bacteria to bacteria but via persistent lifestyles of viruses and are not an exchange phenomenon performed by bacteria.

As new research indicates the agents of horizontal gene transfer are plasmids, retroplasmids, bacteriophages and transposons. They effect DNA recombinations and act in all prokaryotes. DNA movement is achieved through transformation, conjugation and transduction. Transformation is the transfer of DNA between related bacteria mediated by encoded proteins. Conjugation is performed by conjugative plasmids which are independently replicating genetic elements. These elements code for proteins which facilitate their own transfer (Frost et al. 2005). Transduction is a DNA transfer mediated by phages which can package host DNA in their capsid and inject it into a new host followed by integration into the host genome (Holmes et al. 2003). Phages, plasmids, retroplasmids and transposons therefore played a crucial role in bacteria evolution (Chen et al. 2005). Bacteria are the most genetically adaptable organisms with enormous capabilities to react appropriately to extreme changes of their ecological habitats. This does not stem from their high reproductive rates but from their great ability to acquire DNA segments by plasmids, bacteriophages and transposons which transport complete and complex sets of genes from external sources (Shapiro 2007).

When we consider the age of the ocean and the dense abundance of bacterial and viral life in it, then we can say that the possibility of genetic arrangements, rearrangements and exchange does not need long time periods to create the basics of the complexity of life, because the exchange rate is of astronomical order. If we imagine that 1 ml of seawater contains one million bacteria and ten times more viral sequences it can be determined that 10^{31} bacteriophages infect 10^{24} bacteria per second (Tettelin et al. 2005). Since the beginning of life this behavioural pattern has been an ongoing process. The enormous viral genetic diversity in the ocean seems to have established pathways for the integration of complete and complex genetic data sets into host genomes, e.g. acquisition of complex new phenotypes via a prophage can include the acquisition of more than 100 new genes in a single genome editing event (Ryan 2006).

Owing to the virus-induced genomic plasticity of bacteria they are an ideal global biotic matrix to evolve and develop varieties in genome editing, i.e. competent content arrangement of bacterial gene word order coherent with its regulation network. Bacteria are the smallest living organisms with relatively simple genomic structures where the competitive situation between an abundance of viral infective elements leads to the adaptation of lytic viruses to temperate viruses integrated as plasmids in cytoplasma and even persistent viruses integrated in the host genome. The viral competences can develop in this global bacterial habitat as the bacterial species due to their immense genetic flux between viral colonization events and immunity reactions such as restriction/modification (Hambly and Suttle 2005; Kulakauskas et al.1995).

The highly conserved genome edited functions such as replication, transcription, translation, recombination and all the substeps evolved primarily in the competitive situation between viral competences to colonize a host and to ward off competing parasites. This includes that biotic self and non-self recognition functions as we know it from diverse immunity systems are also of viral origin, i.e. the integration and all genetic/genomic modification steps that what we call natural genome editing are of viral origin. Therefore the immense importance of horizontal gene transfer for bacterial species evolution, diversity and competences is derived from viral genome editing competences and is, in most cases, infection induced by persistent non-lytic viruses (Villarreal 1999; Frost et al. 2005). As phylogenetic analyses demonstrate, the main protein enzymes for natural genome editing are viral inventions and not of cellular origin (Villarreal 2004, 2005). Also, the origin of eukaryotic nucleus was thought to be an ancient prokaryote but phylogenetic analyses show that its ancestor was a large DNA virus (Takemura 2001; Bell 2001, 2006). Interestingly, the early genetic invention of capsid proteins detected in viruses infecting archaea seems also to be of viral origin and of common ancestry to eukaryotic and bacterial viruses (Nandhagopal et al. 2002; Rice et al. 2004; Khayat et al. 2005).

6.6 Conclusion

For a long time bacteria have been assumed to be the most primitive organisms and consequently investigated as single-cell individuals determined by mechanistic input-output reactions. Now this picture has changed radically. Today we know that bacteria are parts of bacterial communities which interact in a highly sophisticated manner. The medium of every bacterial interaction is communication, i.e. sign-mediated. A wide range of chemical molecules serve as signs through which bacterial communities exchange information and act in reaching a 'quorum' which is the starting-point for decision-making: one of many different behavioural patterns will thereby be organised, such as biofilm organisation, bioluminescence, virulence or sporulation. Quorum-sensing is not only chemotaxis, but includes interpretation, which means that the incoming signals are measured on the background memory of the species-colony in their real life world. The interpretation before decision-making, coordination and organisation, such as fruiting body formation and cooperative hierarchical organisation, is context-dependent (Fig. 6.5).

Bacteria, which in former times were viewed as lower life-forms, have now been recognised as masters of monitoring, computing, interpretation, coordination and organisation. Bacterial communicative competences are sign-mediated interactions between the same or related species, but also between non-related species according to different situational contexts (pragmatic level of analyses) and the coherent combinatorial patterns of signals according to the molecular syntax (syntactic level of analyses), both determining the content of the messages (semantic level of analyses), the meaning of signalling molecules for a bacterial community which shares a common background memory and a competence for culture-dependent interpretation competence which is an advantage for adaptational purposes.

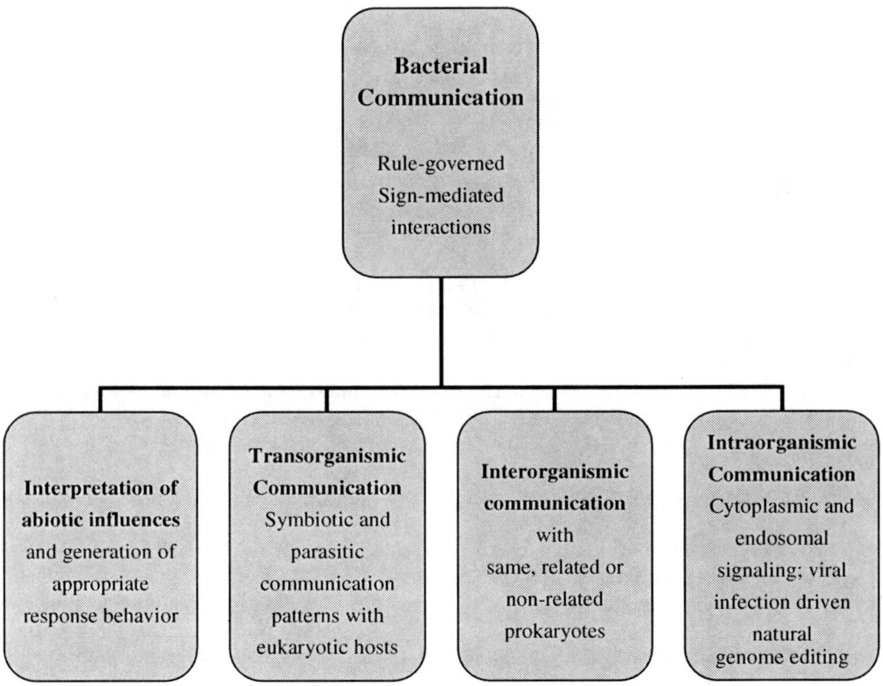

Fig. 6.5 Levels of bacterial communication

References

Bais HP, Park SW, Weir TL et al. (2003) How plants communicate using the underground information superhighway. Trends Plant Sci 9:26–32

Bassler BL (1999) How bacteria talk to each other: Regulation of gene expression by quorum sensing. Curr Opin Microbiol 2:582–587

Bassler BL (2002) Small talk: cell-to-cell communication in bacteria. Cell 109:421–424

Bassler BL, Losick R (2006) Bacterially speaking. Cell 125:237–246

Batut J, Andersson GE, O'Callaghan D (2004) The evolution of chronic infection strategies in the a-proteobacteria. Nat Rev Microbiol 2:933–945

Bauer WD, Robinson JB (2002) Disruption of bacterial quorum sensing by other organisms. Curr Opin Biotech 13:234–237

Bell JL (2001) Viral eukaryogenesis: was the ancestor of the nucleus a complex DNA virus? J Mol Evol 53:251–256

Bell PJL (2006) Sex and the eukaryotic cell cycle is consistent with a viral ancestry for the eukaryotic nucleus. J Theor Biol 243:54–63

Ben Jacob E (2003) Bacterial self-organization: co-enhancement of complexification and adaptability in a dynamic environment. Phil Trans A Mathem Phys Eng Sci 361:1283–1312

Ben Jacob E, Becker I, Shapira Y et al. (2004) Bacterial linguistic communication and social intelligence. Trends Microbiol 12:366–372

Ben Jacob E, Levine H (2006) Self-engineering capabilities of bacteria. J Roy Soc Interf 3:197–214

Berg OG, Kurland CG (2002) Evolution of microbial genomes: sequence acquisition and loss. Mol Biol Evol 19:2265–2276

Blech J (2000) Leben auf dem Menschen. Die Geschichte unserer Besiedler. Rowohlt Taschenbuch Verlag, Hamburg

Bordenstein SR, Reznikoff WS (2005) Mobile DNA in obligate intracellular bacteria. Nat Rev Microbiol 3:688–699

Brown JR (2003) Ancient horizontal gene transfer. Nat Rev Genet 4:122–132

Camara M, Williams P, Hardman A (2002) Controlling infection by tuning in and tuning down the volume of bacterial small-talk. Lancet Infect Dis 2:667–676

Chen I, Christie PJ, Dubnau D (2005) The ins and outs of DNA transfer in bacteria. Science 310:1456–1460

Crespi BJ (2001). The evolution of social behavior in microorganisms. Trends Ecol Evol 16: 178–183

Daniels R, Vanderleyden J, Michiels J (2004) Quorum sensing and swarming migration in bacteria. FEMS Microbiol Rev 28:261–289

Davis DG, Parsek MR, Pearson JP et al. (1998) The involvement of cell-to-cell signals in the development of a bacterial biofilm. Science 280:295–298

Defeu-Soufo JH, Graumann PL (2004) Dynamic movement of actin-like proteins within bacterial cells. EMBO Rep 5:789–794

Dunn AK, Handelsman J (2002) Toward an understanding of microbial communities through analysis of communication networks. Antonie van Leeuwenhoeck 81:565–574

Federle MJ, Bassler BL (2003) Interspecies communication in bacteria. J Clin Invest 112: 1291–1299

Forterre P (2002) The origin of DNA genomes and DNA replication proteins. Curr Opin Microbiol 5:525–532

Forterre P (2005) The two ages of the RNA world, and the transition to the DNA world: a story of viruses and cells. Biochimie 87:793–803

Forterre P (2006) The origin of viruses and their possible roles in major evolutionary transitions. Virus Res 117:5–16

Frost LS, Leplae R, Summers AO et al. (2005) Mobile genetic elements: the agents of open source evolution. Nat Rev Microbiol 3:722–732

Fuqua C, Greenberg EP (2002) Listening in on bacteria: acylhomoserine lactone signalling. Nat Rev Mol Cell Biol 3:685–695

Fuqua C, Winans SC, Greenberg EP (1996) Census and consensus in bacterial ecosystems: the LuxR-LuxI family of quorum-sensing transcriptional regulators. Ann Rev Microbiol 50: 727–751

Gitai Z (2005) The new bacterial cell biolgy: moving parts and subcellular architecture. Cell 120:577–586

Gogarten JP, Townsend JP (2005) Horizontal gene transfer, genome innovation and evolution. Nat Rev Microbiol 3:679–687

Graumann PL (2004) Cytoskeletal elements in bacteria. Curr Opin Microbiol 7:565–571

Graumann PL, Defeu-Soufo HJ (2004) An intracellular actin motor in bacteria? Bioessays 26:1209–1216

Gray KM (1997) Intercellular communication and group behavior in bacteria. Trends Microbiol 5:184–188

Greenberg EP (2003) Bacterial communication: tiny teamwork. Nature 424:134

Guerrero R, Berlanga M (2007) The hidden side of the prokaryotic cell: rediscovering the microbial world. Intern Microbiol 10:157–168

Hambly E, Suttle CA (2005) The viriosphere, diversity, and genetic exchange within phage communities. Curr Opin Microbiol 8:444–450

Hayashi M (2005) Plastid proteins crucial for symbiotic fungal and bacterial entry into plant roots. Nature 433:527–531

Henke JM, Bassler BL (2004) Bacterial social engagements. Trends Cell Biol 14:648–656

Holmes AJ, Gillings MR, Nield BS et al. (2003) The gene cassette metagenome is a basic resource for bacterial genome evolution. Environ Microbiol 5:383–394

Imaizumi-Anraku H, Takeda N, Cherpentier M et al. (2005) Plastid proteins crucial for symbiotic fungal and bacterial entry into plant roots. Nature 433:527–531

Jain R, Rivera MC, Lake JA (1999) Horizontal gene transfer among genomes: The complexity hypothesis. Evolution 96:3801–3806

Jain R, Rivera MC, Moore JE et al. (2003) Horizontal gene transfer accelerates genome innovation and evolution. Mol Biol Evol 20:1598–1602

Kaiser D, Losick R (1993) How and why bacteria talk to each other. Cell 73:873–885

Kaiser D, Welch R (2004) Dynamics of fruiting body morphogenesis. J Bacteriol 186:919–927

Keller L, Surette MG (2006) Communication in bacteria: an ecological and evolutionary perspective. Nat Rev Microbiol 4:249–258

Khayat R, Tang L, Larson ET et al. (2005) Structure of an archaeal virus capsid protein reveals a common ancestry to eukaryotic and bacterial viruses. Proc Natl Acad Sci U S A 102: 18944–18949

Kiers ET, Rousseau RA, West SA et al. (2003) Host sanctions and the legume-Rhizobium mutualism. Nature 425:78–81

Kohlenbrander PE, Andersen RN, Blehert DS et al. (2002) Communication among oral bacteria. Microbiol Mol Biol Rev 66:486–505

Kohlenbrander PE, Egland PG, Diaz PI et al. (2005) Genome-genome interactions: bacterial communities in initial dental plaque. Trends Microbiol 13:11–15

Konaklieva MI, Plotkin BJ (2006) Chemical communication – do we have a quorum? Mini Rev Med Chem 6:817–825

Kreft JU (2004) Biofilms promote altruism. Microbiol 150:2751–2760

Kroes I, Lepp PW, Relman DA (1999) Bacterial diversity within the human subgingival crevice. Proc Natl Acad Sci U S A 96:14547–14552

Kulakauskas S, Lubys A, Ehrlich SD (1995) DNA restriction-modification systems mediate plasmid maintenance. J Bacteriol 177:3451–3454

Kunin V, Goldovsky L, Darzentas N et al. (2005) The net of life: Reconstructing the microbial phylogenetic network. Genome Res 15:954–959

Lenz DH, Mok KC, Lilley BN et al. (2004) The small RNA Chaperone Hfq and multiple small RNAs control quorum sensing in vibrio harveyi and vibrio cholerae. Cell 118:69–82

Loh J, Pierson EA, Pierson LS 3rd et al. (2002) Quorum sensing in plant associated bacteria. Curr Opin Plant Biol 5:285–290

Manefield M, Turner SL (2002) Quorum sensing in context: out of molecular biology and into microbial ecology. Microbiol 148:3762–3764
Margulis L (1996) Archaeal-eubacterial mergers in the origin of Eukarya: Phylogenetic classification of life. Proc Natl Acad Sci U S A 93:1071–1076
Margulis L (1999) Die andere Evolution. Spektrum Akademischer Verlag, Heidelberg
Margulis L (2004) Serial endosymbiotic theory (SET) and composite individuality. Transition from bacterial to eukaryotic genomes. Microbiol Tod 31:173–174
Margulis L, Sagan D (2002) Acquiring Genomes. A Theory of the Origin of Species. Basic Books, New York
McNab R, Lamont RJ (2003) Microbial dinner-party conversations. The role of LuxS interspecies communication. J Med Microbiol 52:541–545
Moore WEC, Moore LVH (1994) The bacteria of periodontal diseases. Periodontology 2000 5:66–77
Nandhagopal N, Simpson AA, Gurnon JR et al. (2002) The structure and evolution of the major capsid protein of a large, lipid-containing DNA virus. Proc Natl Acad Sci U S A 99: 14758–14763
Pal C, Papp B, Lercher MJ (2005) Adaptive evolution of bacterial metabolic networks by horizontal gene transfer. Nat Genet 37:1372–1375
Pandey DP, Gerdes K (2005) Toxin–antitoxin loci are highly abundant in free-living but lost from host-associated prokaryotes. Nucleic Acids Res 33:966–976
Parsek MR, Greenberg EP (2005) Sociomicrobiology: the connections between quorum sensing and biofilms. Trends Microbiol 13:27–33
Paster BJ, Boches SK, Galvin JL et al. (2001) Bacterial diversity in human subgingival plaque. J Bacteriol 183:3770–3783
Rice SA, Givskov M, Steinberg P et al. (1999) Bacterial signals and antagonists: the interaction between bacteria and higher organisms. J Mol Microbiol Biotechnol 1:23–31
Rice G, Tang L, Stedman K et al. (2004) The structure of a thermophilic archaeal virus shows a double-stranded DNA viral capsid type that spans all domains of life. Proc Natl Acad Sci U S A 101:7716–7720
Rickard AH, Palmer JR, Blehert DS et al. (2006) Autoinducer 2: a concentration-dependent signal for mutualistic bacterial biofilm growth. Mol Microbiol 60:1446–1456
Rivera MC, Lake JA (2004) The ring of life provides evidence for a genome fusion origin of eukaryotes. Nature 431:152–155
Ryan F (2006). Genomic creativity and natural selection: a modern synthesis. Biological J Linn Soc 88:655–672
Sahasrabudhe A, Dewhirst FE (2001) Bacterial diversity in human subgingival plaque. J Bacteriol 183:3770–3783
Samaj J, Baluska F, Voigt B et al. (2004) Endcytosis, actin cytoskeleton, and signalling. Plant Physiol 135:1150–1161
Schauder S, Bassler BL (2001) The languages of bacteria. Gen Dev15:1468–1480
Schauder S, Shokat K, Surette MG et al. (2001) The LuxS family of bacterial autoinducers: biosynthesis of a novel quorum-sensing signal molecule. Mol Microbiol 41:463–476
Shapiro JA (1998) Thinking about bacterial populations as multicellular organisms. Ann Rev Microbiol 52:81–104
Shapiro JA (2007) Bacteria are small but not stupid: cognition, natural genetic engineering and sociobacteriology. Stud Hist Phil Biol Biomed Sci 38:807–819
Sharma A, Sahgal M, Johri BN (2003) Microbial communication in the rhizosphere: operation of quorum sensing. Curr Sci 85:1164–1172
Solomon JM, Grossman AD (1996). Who's competent and when: regulation of natural genetic competence in bacteria. Trends Genet 12:150–155
Sperandio V, Torres AG, Jarvis B et al. (2003) Bacteria-host communication: the language of hormones. Proc Natl Acad Sci U S A 100:8951–8956
Spoering AL, Gilmore MS (2006) Quorum sensing and DNA release in bacterial biofilms. Curr Opin Microbiol 9:133–137

Stanley NR, Lazazzera BA (2004) Environmental signals and regulatory pathways that influence biofilm formation. Mol Microbiol 52:917–924

Sun J, Daniel R, Wagner-Dobler I et al (2004) Is autoinducer-2 a universal signal for interspecies communication: a comperative genomic and phylogenetic analysis of the synthesis and signal transduction pathways. BMC Evol Biol 4:36

Swift S, Bainton NJ, Winson MK (1994) Gram-negative bacterial communication by N-acylhomoserine lactones: a universal language? Trends Microbiol 2:193–198

Taga ME, Bassler BL (2003) Chemical communication among bacteria. Proc Natl Acad Sci U S A 100:14549–14554

Takemura M (2001) Poxviruses and the origin of the eukaryotic nucleus. J Mol Evol 52:419–425

Teplitski M, Robinson JB, Bauer WD (2000) Plants secrete substances that mimic bacterial N acylhomoserine lactone signal activities and affect population density dependent behaviors in associated bacteria. Mol Plant Microbe Interact 13:637–648

Tettelin H, Masignani V, Cieslewicz MJ et al. (2005) Genome analysis of multiple pathogenic isolates of Streptococcus agalactiae: implications for the microbial "pan-genome". Proc Natl Acad Sci U S A 102:13950–13955

Thomas CM, Nielsen KM (2005) Mechanisms of, and barriers to, horizontal gene transfer between bacteria. Nat Rev Microbiol 3:711–721

Velicer GJ, Kroos L, Lenski RE (2000) Developmental cheating in the social bacterium Myxococcus Xanthus. Nature 404:598–601

Velicer GJ (2003) Social strife in the microbial world. Trends Microbiol 11:330–337

Velicer GJ, Yu YT (2003) Evolution and novel cooperative swarming in the bacterium Myxococcus Xanthus. Nature 425:75–78

Villarreal LP (1999) DNA virus contribution to host evolution. In: Domingo E, Webster RG, Holland JF (eds) Origin and Evolution of Viruses. Academic Press, London

Villarreal LP, DeFilippis VR (2000) A hypothesis for DNA viruses as the origin of eukaryotic replication proteins. J Virol 74:7079–7084

Villarreal LP (2004). Can viruses make us humans? Proc Am Phil Soc 148:296–323

Villarreal LP, DeFilippis VR, Gottlieb KA (2000). Acute and persistent viral life strategies and their relationship to emerging diseases. Virology 272:1–6

Villarreal LP (2005) Viruses and the Evolution of Life. American Society for Microbiology Press, Washington

Visick KL, Fuqua C (2005) Decoding microbial chatter: cell-cell communication in bacteria. J Bacteriol 187:5507–5519

Vogel J, Sharma CM (2005) How to find small non-coding RNAs in bacteria. J Biol Chem 386:1219–1238

Voloshin SA, Kaprelyants AS (2004) Cell-cell interactions in bacterial populations. Biochemistry (mosc) 69:1268–1275

Wadhams GH, Armitage JP (2004) Making sense of it all: bacterial chemotaxis. Nat Rev Mol Cell Biol 5:1024–1037

Walker TS, Bais HP, Grotewold E et al. (2003) Root exudation and rhizosphere biology. Plant Physiol132:44–51

Wang LH, He Y, Gao Y et al. (2004) A bacterial cell-cell communication signal with cross-kingdom structural analogues. Mol Microbiol 51:903–912

Waters CM, Bassler BL (2005) Quorum sensing: cell to cell communication in bacteria. Ann Rev Cell Dev Biol 21:319–346

Witzany G (2005) Natural history of life: history of communication logics and dynamics. SEED J 5:27–55

Witzany G (2006a) Natural genome-editing competences of viruses. Acta Biotheor 54:235–253

Witzany G (2006b) The *Logos* of the *Bios* 1. Contributions to the Foundation of a Three-Leveled Biosemiotics. Umweb, Helsinki

Witzany G (2007) The *Logos* of the *Bios* 2. Bio-communication. Umweb, Helsinki

Woese C, Kandler O, Wheelis M (1990) Towards a natural system of organisms: proposal for the domains Archaea, Bacteria, and Eucarya. Proc Natl Acad Sci U S A 87:4576–4579
Wuertz S, Okabe S, Hausner M (2004) Microbial communities and their interactions in biofilm systems: an overview. Water Sci Technol 49:327–336
Xie G, Bonner CA, Song J et al. (2004) Inter-genomic displacement via lateral gene transfer of bacterial trp operons in an overall context of vertical genealogy. MBC Biol 2:15

Chapter 7
Natural Genome Editing Competences of Viruses and Virus-Like Agents

Abstract It is becoming increasingly evident that the driving forces of evolutionary novelty are not randomly derived chance mutations of the genetic text, but a precise genome editing by omnipresent viral agents. These competences integrate the whole toolbox of natural genetic engineering, replication, transcription, translation, genomic imprinting, genomic creativity, and all types of enzymatic inventions and genetic repair patterns. Even the non-coding, repetitive DNA sequences which were interpreted as being ancient remnants of former evolutionary stages are now recognized as being of viral descent and crucial for higher-order regulatory and constitutional functions of protein structural vocabulary.

7.1 Introduction

In this chapter I argue that non randomly derived natural genome editing can be envisioned as (a) combinatorial (syntactic), (b) context-specific (pragmatic) and (c) content-sensitive (semantic) competences of viral agents. These (three-leveled) semiotic competences could explain the emergence of complex new phenotypes in single evolutionary events.

After short descriptions of the non coding regulatory networks, major viral life strategies and pre-cellular viral life three of the major steps in evolution serve as examples: There is growing evidence that natural genome-editing competences of viruses are essential (1) for the evolution of the eukaryotic nucleus (2) the adaptive immune system and (3) the placental mammals.

Significant evolutionary modifications can occur by changing the repetitive elements that format the genome architecture rather than, as has been widely assumed so far, by altering protein-coding sequences. This implies that changes in the regulatory framework may lead to new complex phenotypes without even requiring a modification of the coding sequences.

For decades, non-coding regions of the genome have been ignored or declared as 'junk'-DNA. Over the past decade, however, scientists have realized that these regions incorporate decisive higher-order regulatory functions. Within the human genome sequences are only 3% coding for proteins and 97% non-coding ones. But

G. Witzany, *Biocommunication and Natural Genome Editing*,
DOI 10.1007/978-90-481-3319-2_7,

these 3% of the protein-coding sequences match with those of the mouse by up to 88%. Highly developed genomes among mammals contain a nearly identical protein-coding vocabulary.

While bacterial genomes almost lack such repetitive elements, there is a tenfold increase of such segments in the genome of the eukaryotic yeast cells; in the case of the fruit fly *Drosophila*, the repetitive elements already amount to 10%, while in humans they reach 97% (Villarreal 2004:304). These formatting, repetitive elements are a prerequisite for the decisive DNA-editing processes such as expression, replication, repair and recombination.

7.2 Non-Coding Regulatory Networks

Clearly, mobile sequences such as transposons and retroposons (Volff 2006) and non-coding repetitive elements such as LTRs, SINEs and LINEs (long terminal repeats, short interspersed elements, long interspersed elements) enable far-reaching DNA rearrangement and reorganization (Shapiro 2002; Sternberg 2002; Shapiro and Sternberg 2005; Deepak et al. 2003). Together, they play a decisive role in the evolution of new genomic structures (Shabalina and Spiridonov 2004; Shapiro 2004; Sternberg and Shapiro 2005; Pollard et al. 2006). Depending on the organism's state of development, the varying chromatin markers are, thus capable – through different methylation patterns, histone modifications and alternative splicing – of creating a set of 'multiple protein meanings' (Ast 2005) from one and the same genetic data-set (Turner 2000, 2002; Jenuwein and Allis 2001; Brett et al. 2002; Xu et al. 2002; Jaenisch and Bird 2003; True et al. 2004; Wang et al. 2004). This even characterizes the rise of epigenetics, i.e. the view that phenotypic variations, which are heritable, need not be connected with genetic alterations (Jablonka and Lamb 1989, 2002, 2006; Van de Vijver et al. 2002; Van Speybroeck et al. 2002). The question arises as to how and why the evolution of higher genetic complexity is connected to non-coding DNA, formerly termed 'junk'-DNA?

Although we have known for several decades that the unbelievable diversity of enzyme proteins is a practical tool for DNA editing processes, it was unclear by which rules or higher-order regulations they are governed (Witzany 1995, 2000, 2005). It subsequently became clear that higher-order regulations such as co-suppression, suppression of transposition, position effect variegation, several start- and stop-signals, RNA interference, imprinting, chromosomal methylation, transvection and transcriptional and post-transcriptional gene silencing are processed by micro-RNAs (Mattick 2001, 2003, 2005; Mattick and Gagen 2001). In 1992 Watson suspected these higher-order regulatory functions to be hidden in chromosomal structures (Watson et al. 1992). Interestingly, researchers analyzing the linguistic features of non-coding DNA sequences draw the right conclusions (Mantegna et al. 1994; Vendrami 2005). They found coherence between coding DNA with similar structural features as artificial scientific languages and non-coding DNA with similar structural features as everyday language.

The genome, with both its coding and non-coding sections, serves as a genetic, epigenetic and computational cell storage medium that involves different kinds of storage techniques. These include short-time storage (dynamic rearrangements, remodifications, rapidly induced alterations for expression and repair), intermediate storage (epigenetic use) including multiple protein-meanings within the same chromosomal structures by different expression modes through different methylation patterns, and long-time storage, i.e. stable inheritable datasets for many generations (Shapiro and Sternberg 2005). This undoubtedly means that the DNA storage medium in its possible functions depends on different means, i.e. interpretations, of the organism inputs on this medium rather than on the inherent information. This enables changing inheritable priorities, whether computational, intermediate, or long-time storage data – or combinatorial patterns among them (Carroll 2005; Jablonka and Lamb 1989, 2006; Ryan 2006).

The repetitive sequences are highly species-specific and more suitable for the determination of species than the coding sequences (Villarreal 2005). Each taxon organizes and formats its genome architecture differently, i.e. regulation of expression, transcription, replication, and translation are species-specific. Within each taxon, these processes, along with the associated gene architecture, must run in a highly coordinated manner so that they do not disturb each other. Only a co-ordination of the individual steps guarantees that these different actions are performed and maintained successfully: DNA sequences must be read at the right site and at the right time. The precise spatio-temporal co-ordinations are essential in order to sustain vital processes; nonetheless, such co-ordinations can also fail, either by sequence-damage (i.e., mutations) or by organism-induced translational, transductional, repairing or other rearrangement disturbances.

Experience has shown that (1) excision, (2) insertion and (3) combination of the genomic codes are the keys in DNA editing and the basis of evolutionary processes. The contemporary term would be 'natural genetic engineering' (Shapiro 2004). Wittgenstein noted that 'to understand a sentence means to understand a language. To understand a language means to be master of a technique' (Wittgenstein 1972). Nonetheless, there is a clear difference between natural and artificial genetic engineering. The former provides ontologically genuine products that are evident in all living beings and life processes, whereas the latter attempts to achieve modifications and improvements by copying the natural genome-editing competences.

7.3 Major Viral Life Strategies

New research has shown that these non-coding repetitive sequences originated primarily from retroviral RNA (Villarreal 2004, 2005). After persistent non-lytic viral infection, important protein coding regions become incurred within the replicating region of the host organism, whereas repetitive sequences are integrated into the non-coding sections (Villarreal et al. 2000; Villarreal and DeFilippis 2000). In this way, thousands of endogenous retroviral sequences have been integrated into the human genome, with 22 independent retroviral families currently having been identified (Villarreal 2004, 2005; Ryan 2006). The quantity of remaining former viral-gene embedding repetitive elements (embracing an enormous genetic diversity) originally accompanied the protein-coding sequences as control- and/or identification segments. Most endogenous retroviruses have been degraded into formerly connected domains, but they can still be recognized by their three genes *gag, pol* and *env* (Gao et al. 2003; Ryan 2004). The *gag* gene encodes structural proteins, *pol* encodes enzymes such as reverse transcriptase and integrase functions and *env* encodes envelope proteins.

At this stage we know roughly 3600 different kinds of viruses and approx. 30,000 viral subtypes (Villarreal 2005:6). The most numerous variety is found among single-stranded RNA viruses, followed by double-stranded DNA viruses, double-stranded RNA viruses and single-stranded DNA viruses (Villarreal 2005:5). This huge variety of viruses and viral components is even more astonishing when considering the fact that 1 ml of sea water contains about 1 million bacteria and that the number of viral components in the same amount of water is approx. 10 times higher (Villarreal 2005:55). However, their ecological niches are not just limited to purely aquatic or terrestrial environments, but also include species-specific tissues that differ significantly from each other (Villarreal 2005:56). As phylogenetic analysis shows, nearly all organisms of all kingdoms have been infected or highly colonized by viruses since the beginning of life.

Viruses can parasitize almost any replication system – even prebiotic ones – and probably emerged well before the appearance of cellular life forms. Viruses store information that not only pertains to (a) replication proteins but also to (b) morphology and (c) phenotypic diversity. Based on this and the results of phylogenetic analyses and comparative genomics, it is possible to establish viral lines of ancestral origin. These lines of origin can also be non-linear because different parts of viruses contain different evolutionary histories (Villarreal 2005:11).

Since viruses with RNA genomes are the only living beings that use RNA as a storage medium, they are considered to be witnesses of an earlier RNA-world, of a time when DNA did not exist yet (Forterre 2001, 2002, 2005, 2006; Koonin et al. 2006). Negatively stranded RNA viruses have genome structures and replication patterns that are dissimilar to all known cell types. As demonstrated by phylogenetic analyses, cellular replicases are related to each other; however, there is no similarity between RNA-viral replicases and those of any known cell types. This proves the existence of negatively stranded viral RNA-replicases even before cellular life came into being (Villarreal 2005:11). DNA viruses, too, do not give any reference to a cellular origin. Phylogenetic analyses point to an older time scale, as DNA-repairing proteins of DNA viruses do not have any counterparts in cellular biota.

For a long time, viruses were not considered a part of the animated world because they were believed to be exclusively parasitic in nature. They were interpreted as causing an acute infection of the host organisms, using the host's cellular machinery to reproduce, and performing their lytic nature only in order to infect other cells. Although this narrative remains valid, it merely represents a special case of viruses that were unable to reach persistent status by a sessile life-style (Villarreal

2004:305). Most viruses, however, are stable, persistent living beings that do not colonize a host organism for simple selfish purposes. Rather, they display varying survival strategies which can differ according to the host they inhabit (Villarreal 2005:7):

- In bacteria we mainly find double-stranded DNA viruses. Similar viral types are also found in algae, but they are absent in fungi and plants.
- Fungal hosts house mainly double-stranded RNA viruses, whereas
- Plants contain predominantly single-stranded RNA viruses.
- Mammals, on the other hand, are colonized mainly by endogenous retroviruses (ERVs).

Viral persistence in host organisms is crucial because they reliably protect the host against similar parasites. Interestingly, competing viruses are in most cases of the same or related species, whereas unrelated viruses do not compete but interact symbiotically (Rossinck 2005).

As for endogenous retrovirus, they confer the host a distinct and unique genetic identity, non-existent before the colonization. This persistence can either be endonucleic or cytoplasmatic. Persistent viruses are capable of continuous or episodic reproduction in the hosts (Villarreal 2005:5). The main direction is not to infect cells for reproduction with lytic consequences and often epidemic viral diseases, but a sessile life-style without harming the host. This includes different meanings of the term fitness. Thus, lytic life-styles are fit if the reproduction rate reaches epidemic dimensions, whereas the persistent life-style is fit if it does not harm the host and wards off competitive viral infections.

For far too long, viruses have only been considered as poisonous and highly infectious parasites. The fact that viruses are silent companions of virtually all organisms and that they play a decisive role in the evolution of the host has been largely ignored.

Gene functions of cellular biota, which undoubtedly are associated with the persistent life strategy of viruses, include (Villarreal 2005:21):

- Immunity (restriction and modification modules, toxic and antitoxic modules);
- Silencing functions/micro-RNAs, (methylation, suppression);
- Recognition functions (replicate expression, receptors, expression factors);
- Immune regulation (signal mediating, heredity, adaptation).

However, the host must be able to inhibit the replication of a persistent virus; i.e., it must prevent it from turning aggressive and exercising its lytic nature. Especially if the persistence does not reach a stable status, the move-countermove interaction between nucleic acid invader and a host genome may be an on-going process which leads to a special kind of immunity, a defensive RNA-silencing against competing viruses (Fire 2005; Buchon and Vaury 2006). This RNAi interactional pattern is documented to be a very old type of innate immune system.

7.4 Examples of Diverse Viral Life Strategies

7.4.1 Virus Escape

Mitochondria and chloroplast precursors take refuge in prokaryotic hosts where they are protected from cyanophagous colonization. Although the host is equipped with restriction and modification competence that can ward off viral colonization, endosymbiotic mitochondria and chloroplasts lack such competences. Such competences might have existed in the past, but they have been relocated to the hosting organism, thereby establishing a type of immune system against cyanophagous infestation, that are omnipresent in oceans (Villarreal 2005:127).

7.4.2 Wall Off

This is closely associated with calcification (encapsulation of the parasites) and provides some mechanical as well as chemical protection. It was initially based on an RNAi defence system found even among simple marine species such as early protostomes and among their relatives such as crustaceans and insects. As in bivalves, this phenomenon is also found among insects: a chitinous or a mineralized barrier is formed around the parasitic intruder. This is a very archaic viral strategy that can still be observed today as a reaction to parasitic infections in mammals and in the form of a chronic inflammatory response in vertebrates (Villarreal 2005:204).

7.4.3 Addiction Module: Reciprocal Interaction

Addiction modules include antagonistic components acquired in order to survive inside a host without damaging it (Villarreal 2005:302). Acquisition of addiction modules can be reconstructed as a massive viral colonization of the host. Addiction modules consist of a set of genes or functions that are harmless to the host and that include both toxin and antitoxin. In most cases the evolutionary aspect for the host organism is an acquired and hereditary immunity function, e.g. the restriction/modification phenotype. These modules probably paved the way for the evolution of new species, so persistent colonization led to permanent integration without being threatened by any other virus (Villarreal 2005:145); e.g. *Dictyostelium* (Villarreal 2005:178, 181), sharks, bryophytes (Villarreal 2005:238) and ferns (Villarreal 2005:228); this process takes place even today. However, the present diversity of these innovative phenotypes is in turn immensely colonized by viruses and, interestingly, by groups of novel viruses that were non-existent before.

7.4.4 Multiplicity Reactivation

Some of the oldest viruses, the phycodnaviruses and cyanophages, are able to repair UV-light induced damages of their genomes and can reproduce even if the host organisms are exposed to lethal doses of radiation. This ability to replicate is only possible due to active participation of virus-specific repair enzymes, which can re-establish the cell's competence to process macromolecules.

In addition, even lethally irradiated virus can often repair itself. The capability to express subsets of genes is retained even if the viruses' genome experiences lethal radiation damage. Since viral fitness can be consortial, a mixed infection with 'dead' virus can provide a functional complementation of expressed genes in which the complementary gene sets are capable of initiating viral replication and compelling recombination of the necessary combinations of defective viral genomes in order to assemble intact viruses. Therefore, viruses are the only living beings capable of meaningfully recombining text fragments of a damaged genome into a fully functional viral genome capable of self-replication (Villarreal 2005:128).

7.4.5 Sexual Isolation

Once a species is massively infested by an acute and lytic pandemic virus, some of the surviving hosts manage to permanently include this viral genome into their own genome, thereby obtaining permanent immunity against such infections. As soon as they engage in intraspecific reproduction, they lethally infect those siblings that lack this additional immunity. This sexually isolates the immunized from the non-infected members. As an immediate result, the genetically altered siblings experience an interruption of their line of ancestry by acquiring a complex immunological phenotype. This enables the immunized species to establish a novel species incapable of mating with the previously related members without killing them (Villarreal 2004:315, 2005:361).

7.5 Pre-Cellular Life: Early RNA- and DNA-Viruses

RNA genomes probably existed before the appearance of DNA genomes (Villarreal 2005:30). Hence, DNA can be considered to be a modification of RNA, yielding ribonocleotide reductase, followed by only two thymidylate synthases.

Initially, there were only two components, RNA molecules and RNA proteins, which constituted a RNA cell that contained only few, individual DNA components. In turn, they were colonized by parasitic RNA that facilitated the scission into two separate lineages – DNA viruses and RNA viruses – whereby a DNA virus was capable of infecting an RNA virus. This model would at least account for the existence of viral encoded DNA transaction proteins, to which no cellular counterpart exists. In addition, this model would explain the existence of two dissimilar DNA-replication systems.

On the one hand, the absence of repair enzymes within the RNA world gave it a more flexible and creative character. On the other hand, this inability is a drawback when a reliable storing capacity is required. Evidently, the presence of such addiction modules within the highly competitive viral RNA-DNA world must have an advantage. This is particularly true when a DNA virus infects an RNA virus, forcing the latter to establish a bi-layered cell-membrane and to encapsulate the genome by a porous nuclear envelope. Doing so still enables replication, repair and recombination (Villarreal 2005).

Even though the DNA cell gained distinct selective advantages, the RNA parasites still had an astonishingly powerful genomic creativity (Ryan 2006); this imparted distinct survival advantages if environmental conditions changed considerably. Today we know that the DNA world, on its own, would not have brought forth such an incredible diversity by natural selection (Gabora 2006) – let alone established the necessary genetic precondition to create such a high degree of complexity. Overall, the genomic innovation of the RNA world complements that of the more conservative and stable DNA world.

Besides these processes, life also resulted from two alternative innovations: the creation of the archaeal domain and that of the eubacterial domain. Again, this would explain why archaeal lipids are so much different from eukaryotic/bacterial lipids. These lipids contradict eubacterial stereochemistry and differ in the backbone architecture of the hydrocarbon chain.

7.6 Origin of the Eukaryotic Nucleus

The origin of the eukaryotic nucleus serves as a first example for the important role of natural genome-editing competences of viruses. Since the introduction of the Serial Endosymbiotic Theory (SET), it is generally accepted that the eukaryotic cell did not result from random mutations, but from the coordinated union of free-living prokaryotes (Margulis 1996, 1999, 2004; Margulis et al. 2000; Margulis and Sagan 2002; Witzany 2006b). Mitochondria and other organelles clearly descended from these micro-organisms (Odintsova and Yurina 2000, 2005) and the assumption was that the eukaryotic nucleus is also probably of archaeal or bacterial descent.

New evidence supports the idea that eukaryotic nuclei originated before the symbiogenetic integration with mitochondria and chloroplasts (Villarreal 2005: 102). In fact, the nucleus has basic properties that are otherwise absent in eukaryotic cells (Bell 2001, 2006).

Prokaryotes do have circular chromosomes with uniform, standardized origins of replication. Their chromosomes are only loosely attached to chromatin proteins and have different control regions that coordinate and terminate DNA replication. All eukaryotic proteins involved in DNA replication differ from those found in prokaryotes. Hence, nuclear properties of eukaryotes are completely different from those of prokaryotes (Villarreal 2004:306). These differences include:

- use of linear chromosomes, with repetitive termination points and several origins for replication,
- transcription and translation separated by nuclear membranes,
- existence of complex nuclear pore structures that actively mediated RNA translocation

All these properties represent complex phenotypes that require complex coordination of numerous protein functions. None of these functions are present in prokaryotes even though they are considered as predecessors of the eukaryotic nucleus.

The eukaryotic nucleus contains three kinds of DNA-dependent RNA polymerases that differ significantly from RNA polymerases of prokaryotes (Villarreal 2005:106). Even the three kinds of splicing group I-introns (DNA transposase, reverse transcriptase and micro-RNAs) are largely inexistent in prokaryotes, although they are present in viruses of prokaryotes. Bacteria never had introns in any of their coding genes. In addition, no single prokaryotic process is known to account for the tasks of membrane disintegration and restoration as observed in eukaryotes. Eukaryotic tubulin consisting of a highly conservative mitotic spindle is absent in all prokaryotic lines of descent, even though it is one of the most important components in eukaryotes. The eukaryotic pore-structure of the nuclear envelope likewise has no counterpart in the prokaryotic world (Villarreal 2005:119).

Viral proteins on the other hand bind both tubulin as well as actin, thereby triggering polymerisation and mobility functions. Viral genes are directly involved in transposition tasks. Interestingly, all viruses mark their genome, their RNA and their proteins with virus-specific methylases, for example via enzymatic reactions known as base methylation (Villarreal 2005:122).

Viruses were long thought to be parasites capable of 'stealing' host-specific properties. In the meantime, viruses are known to have created new genes as a result of their evolutionary line of descent. In the baculoviruses, for example, GenBank database investigations reveal that 80% of their genes are unique to this group and found nowhere else (Villarreal 2005:274). Gene losses have been documented in baculoviruses, but the 12 losses documented therein are opposed by a staggering acquisition of 255 new genes.

Between 1950 and 1980, scientists recognized that the T4 phage-polymerase proteins are much more similar to the eukaryotic DNA polymerase protein than to any prokaryotic polymerase. We now know that the eukaryotic DNA polymerase and the T4 DNA polymerase do have common origins. Indeed, the T4 represents a huge family of viruses that is capable of infecting both Bacteria and Archaea (Villarreal 2004:307). Hence, it is not surprising that T4-DNA polymerases are found in all three domains of life: Archaea, Bacteria and Eukarya.

Algae were among the first higher eukaryotic organisms that had to deal with viruses. Thus, viruses that infected microalgae must have had a large adaptive potential that accompanied the evolutionary pathway and must have included the characteristic of a protonucleus. The phycodnavirus is a case in point: this calls for examining the entire GenBank database for sequences that may be similar to the DNA polymerase of this particular virus (Villarreal 2004:308). Such sequences must include:

- replication polymerases of all higher eukaryotes as well as of all larger eukaryotic DNA-viruses,
- primer polymerases of eukaryotes,
- repair polymerases of both Archaea and Bacteria.

The DNA polymerase of the CSV1 virus is present at the origin of all eukaryotic replication DNA polymerases and is no doubt a precursor of all polymerases involved in replication of the eukaryotic genome. So far no other viral or prokaryotic DNA polymerase that shares these features is known (Villarreal 2004:310).

The membrane-bound separation of transcription and translation is a characteristics of the poxviruses, more specifically of the Vaccina and other DNA viruses. Moreover, these viruses have a very simple pore-structure that has actively been incorporated from the membrane-bound RNA into the host cytoplasm. A similar situation can be documented with the small chromatin proteins and the linear chromosomes along with their repetitive telomere tails that are so characteristic among various cytoplasmic DNA-viruses, TTV1 and phycodna-viruses. Even the highly complex function of tubulin, as an important coordinating element during chromosomal separation of duplicated strands, is present in DNA-viruses with exactly the same set of functions (Villarreal 2004:311).

It became increasingly obvious that all properties of the eukaryotic nucleus are derived from a large, stable and persistent dsDNA virus with linear chromosomes. The current interpretation is that the precursor of the eukaryotic nucleus was indeed a huge membrane-covered DNA virus that persistently colonized a prokaryotic host (Villarreal 2004:311). Therefore, the hosting cell must have lost its cell wall, with the virus incorporating the prokaryotic genes into its pre-nuclear genome, particularly in cases of encoding for metabolism and translation. This virus was most likely non-lytic because it coordinated both its own replication and its transcription genes, it had a double-layered membrane and a tubulin system in order to wrap chromosomes. Its persistence and its reactivation would imply that (a) the process of cell division (nuclear envelope dispersion and reformation), (b) mitotic duplication (doubling of the chromosomes and allocating them to the progeny cells) and (c) the viral DNA corresponds to the sexual reproductive cycle of the host organism.

Interestingly, all these properties can be found in prokaryotic viruses such as cyanophage, archaeal phage, mycobacterial phage and eubacterial phage (Villarreal 2005:112).

Box 1:

Examples of viral genome editing by inventing new genes as shown by Villarreal

RNA, DNA
replicase, polymerase, integrase
DNA repair
restriction / modification
methylation
bilayer nuclear envelope
eukaryotic nucleus

division of transcription and translation
nuclear pores
tubulin-based chromosome duplication
chitin, calcification
linear chromosomes
innate immune system (MHC-Komplex, RNAi)
adaptive immune system
cartilage, bones
skin, dermal glands for poison, mucus and milk
larvae, egg, placenta, flowering plants
viviparous mammals

7.7 Origin of the Adaptive Immunity

The second example for the important role of viral genome-editing competences is the origin of the adaptive immune system. In the case of persistent viruses, the host must actively be able to take control over viral replication. This requires the host to have some kind of defence/immune system (Villarreal 2005: 209). On the other hand, the persistent virus must also be able to wall off similar viruses trying to take advantage of the host. Lower organisms achieve this with so-called addiction modules – sets of genes that impair the host when absent but are of advantage when present in the host (Villarreal 2004:302). Usually, the negative aspects of such gene sets involve toxic substances, while the positive involve the 'antidote'. At any rate, acute and lytic action might be encountered among closely related species, provided that they are persistent to a specific host. Only in such cases is the hosting organism protected. Such virus-based immune modules are complex genetic systems that include interrelated functional sets that can positively or negatively affect the host; certain environmental conditions or colonization by other viruses can alter the current persistent state by becoming acute, lytic and thereby damaging or even killing those that are not yet infected.

Prokaryotes of the archaeal and bacterial domain are the most adaptive individuals on earth. This raises an obvious question: how do these organisms protect themselves against viral agents? The answer involves utilizing restriction/modification enzymes: Restriction enzymes degrade unmodified DNA with modification enzymes acting only during DNA replication. This simple and innate immune system likewise originated from viruses. Prokaryotes colonized by such viruses possess specific immune response capabilities. The host organism acquires new complex immunologic properties once the viral colonization stabilizes (Villarreal 2004:303).

This, however, does not involve horizontal gene transfer (HTG), but rather large-scale viral inventions of new genes. Some find their way through host lineages and become permanently established within the host. Others have to assess the molecular texture of the bacterial genome as well as its evolutionary potential (Villarreal 2004:305).

A major change in evolution occurred during the transition from urochordates (sea squirts/tunicates) to bony fish. Many common tissue types are still evident, although the bony structure required a huge increase in the genome. This was predominantly achieved by retroposons, which originated from endogenous retroviruses. Urochordates, on the other hand, possess only an innate immune system.

7.7.1 The Acquisition of a Complex New Phenotype

Bony fish are the first in this line of descent that possess both an innate and an adaptive immune system (Villarreal 2005:203). The conclusion is that the entire complexity of an adaptive immune system was acquired at once, probably at the very beginning of the vertebrate line of descent. At the same time, evolution simultaneously yielded jawbones, the vertebral spine, and the skull. Concomitantly, new viral families emerged that had never inhabited any of the previous life forms: four different kinds of RNA-virus families, non-defective, autonomous and abundant retroviruses (Villarreal 2005:204). The adaptive immune system therefore represents an interlinked network of proteins that tag inflammation and other acute processes (cytokine and chemokine), and their receptors and signal transmission systems, that stimulate the humoral and cellular antigen-specific immune response pattern. This must have been acquired in a single event because it is monophyletic (Villarreal 2005:205).

While urochordates do not possess any – except for one – of these features, bony fish possess them all. Sharks are among the earliest vertebrates that featured an adaptive immune system (Villarreal 2004:206). Interestingly, hardly any viruses are found in this group (except for a herpes type strain). They have a very rudimentary, adaptive immune system. Sharks are among the first organisms that include both oviparous as well as viviparous species (Villarreal 2005:207).

The reverse side of an adaptive immune system has a drawback: the non-related haploid chromosomal set is no longer rejected. From an immuno-related perspective it is not recognized as being 'alien' – very much in analogy to the more recent development of placental mammals. Nonetheless, a large quantity of cytokines are transcribed within the placenta, i.e. the immune system is highly active, as is the complementary function of the addiction module. Because the mechanisms of self-nonself-identification are extremely vigilant, they protect the forming progeny from external parasitic attacks.

The acquisition of an adaptive immune system is both a punctual and a variable evolutionary event of the animal kingdom. It enabled the expression of highly complex phenotypes. These phenotypes consist of a self-forming and a dynamically adapting genetic system that recognizes 'non-self' elements and can thereby promptly attack and simultaneously prevent fatal auto-aggression. In the context of a manifold self-identification system, this acquisitional gene-set strategy was developed in order to detect novel non-self agents. Once the system recognizes the presence of 'non-self' agents, it responds by developing a new molecular process that involves the generation of genetic diversity and clonal growth of specific cells capable of detecting such non-self agents. This kind of genetic recombination on a genetic level is found only from this point onwards, not in any predecessors (Villarreal 2005:209).

The overall result of such processes is cells that produce new classes of molecules that can (a) bind and suppress non-self agents or (b) enable amoeboid cytotoxic cells to find and neutralize agent-containing cells. Most of these characteristic properties of an adaptive immune system were lacking prior to vertebrate evolution.

Today, we know that tunicates, the precursors of urochordates, possess a polymorphous MHC-like system which is associated with cell-induced, non-adaptive, amoeboid hemolymphic lethal actions (Villarreal 2005:209). This tunicate system, however, lacks a molecular similarity to the vertebrate MHC system. In order to develop an adaptive immune system so common to vertebrates, the tunicate-like system had to acquire an adaptive component that allowed the non-self elements to be detected as alien. Accordingly, an adaptive immune system destroys non-self elements while protecting those that are part of the system.

These related properties are characteristic for addiction modules. Indeed, the adaptive immune system is an elaborated addiction module. Hosts which incorporated it acquired a system with destructive capabilities. This ability to kill is comparable to a potent toxin that lyses any cell exposed to it. Through this self-recognition process the host must be equally able to prevent autolysis, comparable to an antitoxin (Villarreal 2005:210).

Similar to other addiction modules, the (a) lethal, toxic ability of the adaptive immune system is stable and long-lasting, while (b) the ability to express antitoxic characteristics (self-recognition) is only a temporary feature acquired during the development stage of the immune cell. Thus, the adaptive immune system reveals two aspects of an addiction module (toxic and antitoxic) and a varying stability of the toxin with regard to the protecting antitoxin (Villarreal 2005:210).

Many of these individual elements of an adaptive immune system evolved individually early on in evolution. They definitely existed previously in different viral types rather than in cellular organisms.

What is the most important basic gene-function of an adaptive immune system that favors diversity (necessary for recognizing non-self elements), and where did these genes evolve from? The proteins resulting from recombinase activating genes (RAGs) are responsible for DNA adaptations and new combinations (Villarreal 2005:211). They are the driving agents establishing genetic diversity, thereby producing a manifold display of varying surface receptors. From this perspective, RAGs are the crucial starter proteins for the selection of adaptive immunity. Phylogenetic analyses of these RAGs suggest that precursors existed neither in eukaryotic nor prokaryotic genomes. Instead, RAGs are probably closely related to integrase genes of retroviral origin.

7.7.2 Ancestral Origin of an Adaptive Immune System

So far, no specific viral family that can be considered as a precursor of an adaptive immune system has been isolated. No single virus family possesses all the necessary and required properties. The basic ability to acquire RAG functions and their controlling capability is also accompanied by the property of endogenous retroviral colonization of the host genome (Villarreal 2005:212). The above properties point to the acquisition of adaptive immunity rather than a complex punctual genetic event. This acquisition event however, was not the product of an individual genetic parasite but rather the joining of the stable colonization through entire sets of supplementary and defective viral agents (Villarreal 2005:212). These agents must have covered the immediate precursors of cartilaginous fish with their extremely complex addiction module. This enabled both a stable colonization and successfully excluded competing parasites. The host-predecessor was most certainly equipped with a recognition system quite similar to MHC – a system of cytotoxic cells that was colonized and regulated by a complementary set of newly colonizing parasitic agents. This eventually resulted in the creation and evolution of the adaptive immune system (Villarreal 2005:213).

7.8 Evolution of Placental Mammals

The third example for the important role of genome-editing competences of viruses is the origin of placental mammals. The close ties between the human genome and the colonization with repetitive sequences of retroviral origin (LTRs, SINEs, LINEs) become obvious when considering the Y-chromosome (Villarreal 2004:314). It contains only 20–30 coding sequences, approximately 225 genes (Villarreal 2005:338), whereby the majority are non-coding sequences with higher-order regulatory functions. The remainder of endogenous retroviruses is mainly found on the Y-chromosome. Interestingly, the genome-editing competence of Alu elements is evident in that they can change their own gene expression by modifying their own genomic meythylation status (Batzer and Deininger 2002; Ryan 2004)

The most active period of endogenous retroviral transcription occurs during the formation of placental tissue, during growth periods, and when trophoblasts join together (Villarreal 2004:314). Trophoblasts encapsulate the egg, help the egg nest properly, trigger processes that ensure nutrition, and prevent reactive responses by the mother's own immune system. The egg is protected by trophoblasts against an immuno-reactive response of the mother. These characteristics are unknown to monotreme mammals and marsupials. The acquisition of such abilities must have been a remarkable event (Villarreal 2004:314).

In turn, the trophectoderm is a very complex tissue that is, surprisingly, not of maternal origin, but a derivative of the fertilized egg – it even develops before the egg becomes implanted into the uteral lining. Experiments that suppressed expression of endogenous retroviruses inhibited implantation into the uteral lining.

This implies that implantation of the embryo requires transcription of retroviral syncytin-coding genes. In humans, the HERV W *env* gene codes for syncytin (Dupressoir et al. 2005), a molecule used by the host to join trophoblast cells with the tissue that eventually nourishes the embryo (Villarreal 2004:315).

Although these processes have already been known for over 30 years, the purpose of this reaction was unclear: it did not make sense that the evolutionary innovation of placental mammals was tied to the acquisition of a complex set of endogenous retroviruses. Since the trophectoderm is protected by the maternal immune system, it enables further growth into the placenta, thereby modifying blood flow and nutrient supply between mother and embryo. The trophectoderm is associated with extremely high expression rates of ERV genes that result in RNAs as well as in other gene products and retroviral corpuscles. ERVs are highly host-specific and are closely associated with LINEs and SINEs of placental species. Their expression is not suppressed in the trophectoderm.

Once the sex of the totipotent embryo is determined, the high ERV expression rates are stopped and DNA methylation starts functioning (Villarreal 2005:325). Interestingly, ERV competences protect the embryo from the maternal immune system until the embryo's sex is determined. Only then does DNA methylation mediate between growth and development.

There are, however, clear references for evolutive and physiologically relevant qualities. For example, the expression of HERV-3 is boosted because it involves many fetal tissues in humans such as the adrenal cortex, kidney tubules, tongue, heart, liver and central nervous system as well as sebaceous glands of normal skin (Ryan 2006). Thus, important tissues are formed during the fetal stage and are mediated via the presence of human endogenous retroviruses that were expressed during early mitotic divisions. They safeguard not only the formation of the placenta but also of the most important tissues of the fetus. If these retroviral components were removed from our genomes, we would be already extinct.

7.9 Conclusion

Phylogenetic analyses as well as GenBank database comparisons show that most natural genome-editing competences are not of cellular origin but represent original skills of viruses.

Far from being only stochastic results of randomly derived chance combinations as proposed earlier (Eigen et al. 1981; Maynard-Smith 1983; Wächtershäuser 1992; Jortner 2006; Deamer et al. 2006; Szathmary 2006), genome-editing competences are built on precise combinatorial rules (molecular syntax), highly sensitive to interactional contexts (pragmatic rules) that determine differences in the semantic content, e.g. multiple protein meanings generated according to higher-order regulatory functions (Witzany 2000; 2006a). Natural genome editing of viruses depends primarily on their semiotic competences, i.e. their capability to generate and constitute genetic content as a language-like text according to syntactic, pragmatic and semantic rules in coherence with the laws of physics and chemistry. Through these competences the genetic code is used as a kind of innovation-sharing protocol (Vetsigian et al. 2006).

Viruses have two completely different life strategies, which are clearly reflected in their genomes. Accordingly, acute viruses exhibit lytic action and can induce disease and even death, whereas the life strategy of persistent viruses implies compatible interactions with the host. The latter are either integrated into the hosting genome or the cell plasma, and act non-destructively during most life stages of the host. The persistent life-style allows the virus to transmit complex viral phenotypes to the hosting organism. Doing so enables the host to broaden evolutive potentials that can lead to the formation of new species.

The natural genome-editing competences of viruses are most complex in bacteria, in which the complete nucleotide-word-order is largely determined, combined, and recombined by viruses. Hence, the main genomic novelties are found in the prokaryotic domain from where they originally evolved into the higher life forms. Probably all basic enzymatic variations originated therein (Villarreal 2006, personal communication). Massive viral colonization occurred from the very beginning of life, starting with the evolution of Bacteria and Archaea, later on of Protoctista and multicellular Eukaryota. The formation of all kingdoms, their families, genera, and species relies on the effects of viral colonization and results in diversified lineages and ultimately in the evolution of new species.

Increasing complexity and diversity involved genetic innovations, new combinatorial patterns of genetic content, non-coding regulatory networks and modifications of the genomic architecture. Interestingly, increasing complexity correlates with the expansion of non-coding DNA (Taft and Mattick 2004). Combinatorial and rearranging processes in evolutionary and developmental processes occur non-randomly. They need successful and coherent adherence to the rules of molecular syntax (Eigen and Winkler 1975; Witzany 1995). Genetic content within the genomic matrix depends on the situational contexts in which living organisms are involved in vivo (e.g. growth, mating, virulence, stress, etc.). This content can therefore produce multiple protein-meanings, i.e. different semantic contents of the same genetic data-set. This is the prerequisite for epigenetically induced evolutionary and developmental processes. These rule-following processes may fail, with fatal consequences for the organism (Witzany 2006a). If evolutionary processes are intertwined between different species, then complexity is even more evident (Villarreal 2005; Zhang 2006). This indicates an important role of symbiogenetic processes in enhancing genetic, genomic and phenotypic complexity and diversity.

From this perspective, evolutionary history emerges entirely differently than previously thought. This history is not an aggregation of chance mutations of the genetic text and its associated selection. Rather, it is a permanent and competent processing of genetic sequences to acquire previously unknown abilities and to ward off competing parasites via genomic innovation. This explains why scientists had difficulties understanding evolutive patterns before comparative genomics was established: the thesis of natural genome-editing competences of viruses would merely have been another curious hypothesis. Contemporary sequence analyses and intercomparisons underline the crucial evolutionary functions of viruses and their genome-editing competences for all life forms.

References

Ast G (2005) The alternative genome. Sci Am 292:58–65
Batzer MA, Deininger DL (2002) ALU repeats and human genomic diversity. Nat Rev Genet 3:370–380
Bell PJL (2001) Viral eukaryogenesis: was the ancestor of the nucleus a complex DNA virus? J Mol Evol 53:251–256
Bell PJL (2006) Sex and the eukaryotioc cell cycle is consistent with a viral ancestry for the eukaryotic nucleus. J Theor Biol 243:54–63
Brett D, Pospisil H, Valcarcel J et al. (2002) Alternative splicing and genome complexity. Nat Genet 30:29–30
Buchton N, Vaury C (2006) RNAi: a defensive RNA-silencing against viruses and transposable elements. Heredity 96:195–202
Carroll SB (2005) Evolution at two levels: on genes and form. PLoS Biol 3(7): e245.
Deamer D, Singaram S, Rajamani S et al. (2006) Self-assembly processes in the prebiotic environment. Phil Trans R Soc 361:1809–1818
Deepak G, Majumder PP, Rao CB et al. (2003) Nonrandom distribution of alu elements in genes of various functional categories: insight from analysis of human chromosomes 21 and 22. Mol Biol Evol 20:1420–1424
Dupressoir A, Marceau G, Vernochet C et al. (2005) Syncytin-A and syncytin-B, two fusogenic placenta-specific murine envelope genes of retroviral origin conserved in Muridae. Proc Natl Acad Sci USA 102:725–730
Eigen M, Schuster P, Gardiner W et al. (1981) The origin of genetic information. Sci Am 244: 78–94
Eigen M, Winkler R (1975) Das Spiel – Naturgesetze steuern den Zufall. Piper, München
Fire A (2005) Nucleic acid structure and intracellular immunity: some recent ideas from the world of RNAi. Quart Rev Biophys 38:303–309
Forterre P (2001) Genomics and early cellular evolution. The origin of the DNA world. Comptes rendus de l'Académie des sciences. Série 3, Sciences de la vie 324:1067–1076
Forterre P (2002) The origin of DNA genomes and DNA replication proteins. Curr Opin Microbiol 5:525–532
Forterre P (2005) The two ages of the RNA world, and the transition to the DNA world: a story of viruses and cells. Biochimie 87:793–803
Forterre P (2006) The origin of viruses and their possible roles in major evolutionary transitions. Virus Res 117:5–16
Gabora L (2006) Self-other organization: why early life did not evolve through natural selection. J Theor Biol 241:443–450
Gao X, Havecker ER, Baranov PV et al. (2003) Translational recoding signals between *gag* and *pol* in diverse LTR retrotransposons. RNA 9:1422–1430
Jablonka E, Lamb MJ (1989) The inheritance of acquired epigenetic variations. J Theor Biol 139:69–83
Jablonka E, Lamb MJ (2002) The changing concept of epigenetics. In: VanSpeybroeck L, Van de Vijver G, Waele DD (eds) From Epigenesis to Epigenetics. The genome in context. Ann NY Acad Sci 981:82–96
Jablonka E, Lamb MJ (2006) The evolution of information in the major transitions. J Theor Biol 239:236–246
Jaenisch R, Bird A (2003) Epigenetic regulation of gene expression: how the genome integrates intrinsic and environmental signals. Nat Genet 33:245–254

Jenuwein T, Allis CD (2001) Translating the histone code. Science 293:1074–1080
Jortner J (2006) Conditions for the emergence of life on the early earth: summary and reflections. Phil Trans R Soc 361:1877–1891
Koonin EV, Senkevich TG, Dolja VV (2006) The ancient virus world and evolution of cells. Biology Direct 1:29.
Mantegna RN, Buldyrev SV, Goldberger AL et al. (1994) Linguistic features of noncoding DNA sequences. Phys Rev Let 73:3169–3172
Margulis L (1996) Archaeal-eubacterial mergers in the origin of Eukarya: Phylogenetic classification of life. Proc Natl Acad Sci USA 93:1071–1076
Margulis L (1999) Die andere Evolution. Spektrum Akademischer Verlag, Heidelberg
Margulis L (2004) Serial endosymbiotic theory (SET) and composite individuality. Transition from bacterial to eukaryotic genomes. Microbiol Tod 31:173–174
Margulis L, Dolan MF, Guerrero R (2000) The chimeric eukaryote: origin of the nucleus from the karyomastigont in amitochondriate protists. Proc Natl Acad Sci USA 97:6954–6959
Margulis L, Sagan D (2002) Acquiring Genomes. A Theory of the Origin of Species. Basic Books, New York
Mattick JS (2001). Non-coding RNAs: the architects of eukaryotic complexity. EMBO Rep 2: 986–991
Mattick JS (2003) Challenging the dogma: the hidden layer of noncoding RNAs in complex organisms. BioEssays 25:930–939
Mattick JS (2005) Das verkannte genom-programm. Spektrum der Wissenschaft 3:62–69
Mattick JS, Gagen MJ (2001) The evolution of controlled multitasked gene networks: the role of introns and other noncoding RNAs in the development of complex organisms. Mol Biol Evol 18:1611–1630
Maynard Smith J (1983) Models of evolution. Proc R Soc Biol Sci 219:315–325
Odintsova MS, Yurina NP (2000) RNA editing in plant chlorplasts and mitochondria. Fisiologia Rastenij 37:307–320
Odintsova MS, Yurina NP (2005) Genomics and evolution of cellular organelles. Russ J Genet 41:957–967
Pollard KS, Salama SR, Lambert N et al. (2006) An RNA gene expressed during cortical development evolved rapidly in humans. Nature 443: 167–172
Roossinck MJ (2005) Symbiosis versus competition in plant virus evolution. Nat Rev Microbiol 3:917–924
Ryan FP (2004) Human endogenous retroviruses in health and disease: a symbiotic perspective. J R Soc Med 97:560–565
Ryan FP (2006) Genomic creativity and natural selection. A modern synthesis. Biol J Linn Soc 88:655–672
Shabalina SA, Spiridonov NA (2004) The mammalian transcriptome and the function of non-coding DNA sequences. Genome Biol 5: 105e.
Shapiro JA (2002) Genome organization and reorganization in evolution: formatting for computation and function. In: Van Speybroeck L, Van de Vijver G, Waele DD (eds) From Epigenesis to Epigenetics. The genome in context. Ann NY Acad Sci 981:111–134
Shapiro JA (2004) A 21st century view of evolution: genome system architecture, repetitive DNA, and natural genetic engineering. Gene 345:91–100
Shapiro JA, Sternberg R (2005) Why repetitive DNA is essential to genome function. Biol Rev 80:1–24
Sternberg R (2002) On the roles of repetitive DNA elements in the context of a unified genomic-epigenetic system. In: Van Speybroeck L, Van de Vijver G, Waele DD (eds) From Epigenesis to Epigenetics. The genome in context. Ann NY Acad Sci 981:154–188
Sternberg R, Shapiro JA (2005) How repeated retroelements format genome function. Cytogenet Genome Res 110:108–116
Szathmary E (2006) The origin of replicators and reproducers. Phil Transact R Soc 361:1761–1776
Taft RJ, Mattick JS (2004) Increasing biological complexity is positively correlated with the relative genome-wide expansion of non-protein-coding DNA sequences. Genome Biol 5:1

True H, Berlin I, Lindquist SL (2004) Epigenetic regulation of translation reveals hidden genetic variation to produce complex traits. Nature 431:184–187

Turner BM (2000) Histone acetylation and an epigenetic code. BioEssays 22:836–845

Van Speybroeck L, Van de Vijver G, De Waele D. (2002). Preface. In: Van Speybroeck L, Van de Vijver G, Waele DD (eds) From Epigenesis to Epigenetics. The genome in context. Ann NY Acad Sci 981:vii

Van de Vijver G, Van Speybroeck L, De Waele D (2002) Epigentics: a challenge for genetics, evolution and development. In: Van Speybroeck L, Van de Vijver G, Waele DD (eds) From Epigenesis to Epigenetics. The genome in context. Ann NY Acad Sci 981:1–6

Vendrami D (2005) Noncoding DNA and the teem theory of inheritance, emotions and innate behaviour. Med Hypothes 64:512–519

Vetsigian K, Woese C, Goldenfeld N (2006) Collective evolution and the genetic code. Proc Natl Acad Sci USA 103:10696–10701

Villarreal LP (2004) Can viruses make us humans? Proc Am Phil Soc 148:296–323

Villarreal LP (2005) Viruses and the Evolution of Life. American Society for Microbiology Press, Washington

Villarreal LP, DeFilippis VR (2000) A hypothesis for DNA viruses as the origin of eukaryotic replication proteins. J Virol 74:7079–7084

Villarreal LP, DeFilippis VR, Gottlieb KA (2000) Acute and persistent viral life strategies and their relationship to emerging diseases. Virology 272:1–6

Volff JN (2006) Turning junk into gold: domestication of transposable elements and the creation of new genes in eukaryotes. BioEssays 28:913–922

Wang Y, Fischle W, Cheung W et al. (2004) Beyond the double helix: writing and reading the histone code. In: Bock G, Goode J (eds) Reversible Protein Acetylation. Novartis Foundation 2004, 259:3–17

Watson JD, Witkowski JA, Gilma M et al. (1992) Recombinant DNA, 2nd ed. Scientific American Books, New York

Wittgenstein L (1972) Philosophical Investigations. Basil & Blackwell, Oxford

Witzany G (1995) From the 'logic of the molecular syntax' to molecular pragmatism. Explanatory deficits in Manfred Eigen's concept of language and communication. Evol Cog 1:148–168

Witzany G (2000) Life: The Communicative Structure. A New Philosophy of Biology. Libri Books on Demand, Norderstedt

Witzany G (2005) Natural history of life: history of communication logics and dynamics. SEED J 5:27–55

Witzany G (2006a) The *Logos* of the *Bios* 1. Contributions to the Foundation of a Three-Leveled Biosemiotics. Umweb, Helsinki

Witzany G (2006b) Serial endosymbiotic theory (SET): The biosemiotic update. Acta Biotheor 54:103–117

Wächtershäuser G (1992) Groundworks for an evolutionary biochemistry: the iron-sulphur world. Progr Biophys Mol Biol 58:85–201

Xu Q, Modrek B, Lee C (2002) Genome-wide detection of tissue-specific alternative splicing in the human transcriptome. Nucleic Acids Res 30:3754–3766

Zhang HY (2006) The evolution of genomes and language. EMBO Rep 7:248–249

Turner BM (2002). Cellular memory and the histone code. Cell 111:285–291

Chapter 8
How Bacteria Escaped Selection Pressure of the Early RNA-World

Abstract From a biocommunicative perspective it is suggested that in the beginning of life there were single-stranded, unencapsulated RNA molecules with an aptitude for replicating themselves by copying. The crucial shift was the invention of coding capabilities, because this dramatically differs from pure copying in a function which assembles a molecular syntax, pragmatics (context) and semantics (content) in parallel. Because coded nucleotide sequences differ from a random mixture of nucleotides, in that coded sequences imply a molecular grammar whereas a random mixture of nucleotides does not, it is necessary to propose coding agents which are competent to differentiate self from non-self structures and colonize competing agents by new sequence inventions which would serve as advantage in both increasing genetic sequence-complexity and immune function against competing agents. These pre-requisites are fulfilled by consortial interacting ribozymes. i.e. last universal common pre-viruses. The fittest of them escaped by invention of consortial addiction modules and protein membranes

8.1 Introduction

For a long time bacteria have been considered to be the forerunners of the eukaryotic superkingdom. Although the evolution of eukaryotes did not occur by random mutations of bacterial genomes but by integration and natural genetic engineering of former free-living prokaryotes (Witzany 2006a) the key features of the eukaryotic nucleus have less in common with prokaryotic competences than with some double-stranded (ds)DNA viruses (Witzany 2006b). The textbook conviction of the early 21st century on the evolutionary history of eukaryotes was that an ancient prokaryotic cell was colonised by a large dsDNA virus and afterwards by mitochondria-like and chloroplast- like bacteria which together built the first eukaryotic cell. This scenario makes sense from a cytological perspective, because prokaryotes are much simpler than eukaryotes. From the perspective of an early RNA world, however, this picture could be completed.

G. Witzany, *Biocommunication and Natural Genome Editing*,
DOI 10.1007/978-90-481-3319-2_8, © Springer Science+Business Media B.V. 2010

8.2 From Pre-Cellular RNA-Copying to RNA-Coding

A 'virus-first' scenario from a biocommunicative perspective would look like this. In the beginning there were single-stranded, unencapsulated RNA molecules with an aptitude for replicating themselves by copying via Watson-Crick basepairing. The crucial shift was the invention of coding capabilities, because this dramatically differs from pure copying in a function which assembles a molecular syntax, pragmatics and semantics in parallel. Because coded nucleotide sequences differ from a purely random mixture of nucleotides, in that coded sequences imply a molecular grammar whereas a random mixture of nucleotides does not, they built the real precondition to generate more complex structures with multiple functions to form dsRNA genomes in a pre-DNA world out of the shortest functioning RNA sequences (Flores et al. 2004).

If we term these pre-cellular RNA-coding replicators as viruses then ssRNA viruses evolved into dsRNA viruses. Via a reverse transcriptase function present in an RNA-dependent RNA-polymerase (Chiang et al. 1994; Chiang and Lambowitz 1997), these dsRNA viruses evolved later on into dsDNA viruses. RNA-dependent RNA polymerases are core competences of RNA viruses (Villarreal 2009).

At this stage, reverse transcriptases - the only competence which is encoded by all of the known retroagents - must still have existed. Even retroagents are ssRNAs. On an evolutionary timescale retroagents emerged after the coding and RNA-editing (RNA sequence order determining) competences of ssRNAs. These early retroagents shared a kind of immune function against colonisation by ssRNA-parasites and -hyperparasites. The reverse transcriptase in the retrosyntax used by retroagents acts as a polymerase, whereas polymerases from this perspective function as reverse transcriptases. The (retro)direction of sequence-reading serves as an efficient barrier against polymerase-using genetic parasites.

If we wished to reconstruct the evolutionary emergence of the stable DNA storage medium, we would probably have to look at the highly conserved integration of nucleotide sequence-reading by polymerases and its complementary counterpart the retrodirection-reading by reverse transcriptases. A current view on the result of this evolutionary integration is possible if we look at the tools by which one of the most fundamental steps in living nature is brought about, i.e. DNA replication and the different processes of leading and lagging strand replication. Whereas leading strand synthesis is a continuous processing by DNA polymerase III, lagging strand synthesis occurs in several highly conserved coordinated steps including not merely DNA polymerase III but RNA Primase also.

Retroagents are the only competences that process DNA and therefore DNA viruses from RNA sequences, and they could act as persistent endogenous retroviruses coding within DNA structures such as DNA viruses or, later on, eukaryotic genomes. Endogenized nucleotide sequences of retroagents we can identify not solely by their three essential genes of *pol*, *gag* and *env* but by non-coding repetitive elements also (see Chapter 7) which may be mobile (retroposons) or highly conserved and non-mobile such as telomeres and centromeres (see Chapter 9).

Only retroelements with their inherent encoded reverse transcriptases are capable of transcribing an RNA Code into a DNA Code.

Now the stable DNA of dsDNA viruses was advantageous for colonising the unstable nucleotide word order in the genomic contents of RNA viruses. In parallel, DNA of dsDNA viruses served as an appropriate habitat for infection events by retroid agents. By holding these colonisation interrelations in a non-lytic but persistent inheritable status, infection forced dsDNA viruses to encapsulate the genome in a porous bi-layered nuclear envelope. Currently this could be a coherent explanation for the three remaining cellular DNA replication competences, i.e. the beginning of cellular life would have been completely entangled with three population-like genetic lineages, similar to those suggested by Forterre and co-authors (Forterre 2005). Additionally, DNA repetitive sequences are a strong indicator of persistent colonisation events by retroagents bearing enormous potential in genetic regulations and natural genetic engineering.

These steps from (i) ssRNA and (ii) ssRetroids/reverse transcriptase to dsRNA, dsRNA to dsDNA, are hallmarks in the evolution of life from a prebiotic assembly of ribonucleotides into functional agents with simple nucleotide grammar-editing abilities. These agents also, however, had to include a self/non-self differentiation capability, being able to ward off competing agents by using an ancient immune function similar to siRNA and its RNAi pathways (Obbard et al. 2009). In parallel this would have been an advantage in colonising abundant RNA copies that lacked this capability. These simple pre-cellular RNA copies are an abundant and freely available resource for RNA agents competent in RNA editing.

At this stage we are at the crossroads of life and non-living matter, between chemical reactions such as the sole copying of random mixtures of nucleotides by Watson-Crick base-pairing and on the other side genetic text-editing competences represented by small RNA agents which can identify, target, silence, edit, repair and encode language-like nucleotide grammar. With their inherent self/non-self recognising capability they would be able to colonise prebiotic RNA copies and generate immune function against competing coding agents. This whole scenario could have happened within the high density of a viral soup of archaic oceans.

8.3 Communal Evolution: From LUCA to LUCAs

For a long time the first living cell has been imagined as a rather selfish single agent, known as the last universal common ancestor (LUCA), which was the forerunner of all later evolutionary steps into the three domains of life (Forterre et al. 2005). According to our biocommunicative scenario the pre-cellular agents that were competent in RNA-, DNA- and Retro-editing were population-like nucleic acid sequence-editing species with commonly shared group behavior (self/non-self identification), group identity, and group interpretation.

This would be coherent with linguistic research, which states that (i) every biotic agent that is competent to use and interpret linguistic-like signs needs a community (either as a gene-pool and/or interactional) with which it shares these capabilities, i.e. linguistic competences cannot emerge out of isolated individuals (Witzany 2000) and (ii) if linguistic-like competence for editing of nucleic acid languages evolved, the capability to generate not only simple but also complex new sequences would grow exponentially and not arithmetically. This could explain both the great diversity of singlecelled life soon after pre-cellular life-processes started and the common feature of evolutionary processes, i.e. the invention of new genetic data sets as whole sequences, genes or geneblocks.

When the pre-cellular consortium of three different viral - or even subviral - lineages developed a common genetic code, which further on served as a stable DNA storage medium for the evolution of the three cellular domains, I suggest we should be talking about the last universal common ancestors (LUCAs), because one single ancestor couldn't evolve both (i) sequence editing competences (ii) a competence for sign-mediated interaction necessary for coordination of common behavior (group identity) and self/non-self identification (Vetsigian et al. 2006; Villarreal 2009). The linguistic competence to generate and engineer genetic text sequences and the communicative competence to generate and perform sign-mediated interactions cannot be reconstructed as solus-ipse events which could be performed by individuals in principle but only as shared interaction patterns.

8.4 Old but Good: Current Competences from an Ancient World

Interestingly, even today we can look at relics of precellular evolution in both RNA viruses and viroids. Viroids and their monophyletic sister group, satellite RNAs, are short circular ssRNAs, viroids being unencapsulated whereas satellite RNAs are encapsulated. We know that viroids have extreme plasticity in their nucleotide sequences, being the most rapidly evolving biological agents (Diener 1989, 1991, 1993, 1995). Important features of small RNAs such as RNA silencing seem to derive from viroid competences (Wang et al. 2004; Daros et al. 2006; Itaya et al. 2007). The most conserved competences of RNA viruses and viroids are RNA stem-loop structures, which play important roles in priming and replication, with an inherent self/non-self differentiation competence in that they determine RNA replication to viral and not to host RNA molecules (Villarreal 2005).

We now can imagine eukarya-like dsDNA viruses with the ribozymatic function of endonucleases competent in RNA-splicing, excision of introns out of tRNAs (Calvin and Li 2008), integration of retroid DNA (McClure 2000), and its key features, a double membrane, linear chromosomes with telomere ends, intronic elements with regulatory functions (Birgisdottir and Johansen 2005), segregation of transcription and translation and the subviral competences which we find in the ribonucleoprotein structures of pre mRNA, pre-tRNA and pre-rRNA, all processed by small nucleolar (sno)RNAs and small nuclear (sn)RNAs. As we know today, the precursor RNAs are a highly sophisticated network of regulatory agents, each of them with a separate RNA processing pathway. Although in prokaryotes we do not find linear chromosomes with telomere repeats, the ancient nuclear pore complex

(Bapteste et al. 2005) or the highly mobile genetic settlers inherent in introns that are competent in RNA splicing, we do find them in eubacterial and archaeal phages.

In addition, prokaryotes share a circular genome with nearly intron-free genetic syntax, whereas the seemingly evolutionarily later eukaryotes have linear chromosomes with telomere repeats to protect their ends against genetic invaders and genomes that are highly colonised by virus derived agents such as transposons, retroposons and related genetic settlers.

Although the 'error-prone' coding-fidelity of the RNA world at the beginning was an advantage for fast adaptation, the evolutionary target evolved into both the relatively stable DNA configuration (via the reverse transcriptase competence-the only encoded function common to all retroelements) and the resistant protein world necessary in the high temperature environments of archaeal populations. Prokaryotes lack the key features of the early RNA world and therefore they would appear to be specialised fast-adapting single celled organisms that used the advantages of the stable DNA storage medium to code for highly temperature resistant protein structures to protect this storage medium.

Although accelerated ssRNA processing of mRNA, tRNA and rRNAs in linear RNA genomes built core competences for natural genome editing in the early RNA world, those ssRNAs without cellular habitats are extremely thermolabile and could not survive in high temperature environments (Penny and Poole 1999). The lack of RNA correction and repair and the high rate of replication combined with innovation allowed a rate of recombination events $1–10 \times 10^6$ times faster than in DNA genomes (McClure 2000). RNA-based life-forms could evolve millions of times faster than DNA-based systems. This was an advantage for the exploration and invention of new sequence space, i.e. new genomic content with phenotypic competences and functions. In contrast, circular genomes with few higher order regulatory elements (represented by a diversity of genetic parasites present in intron-like genomic habitats) had more advantages in a high temperature environment and could adapt faster because of their ability to exchange selected phenotypes within and between protein coding data-sets, as happens in horizontal gene transfer. So RNA cultures with eukaryote-like RNA-processing seem to predate the evolution of prokaryotes, which adapted to fast changing environmental conditions by reducing their genomic content to a DNA with nearly analog (intron-free) protein-coding data-sets. This could be the evolutionary pathway from ribozymes of the early RNA world to ribonucleoproteins via low complexity RNA-chaperones to a DNA protein-based life (Diener 1989; Penny and Poole 1999; Poole et al. 1998; Poole et al. 1999). That eukarya-like genomes predated prokaryotic genomes is consistent with the existence of telomeres and telomere-like functions in ancient dsDNA viruses that seem to be the ancestors of the eukaryotic nucleus and are not part of prokaryotic genomes, although some are found in persistent bacteriophages (see below).

Multiple small regulatory RNAs also play important roles in the bacterial expression of target genes at the posttranscriptional level. They are immediately available after being transcribed from the non-protein-coding sections of bacterial genomes, unlike protein enzymes which must be translated as well (Masse and Gottesman 2002; Wassarman 2002; Majdalani et al. 2005; Tu and Bassler 2007; Hammer and Bassler 2007).

From the perspective of evolutionary history, bacteria seemed to reduce the predated RNA based metabolism of early eukarya-like genetic content arrangements to become specialised in highly-selective environmental conditions such as high temperature and/or fast changing nutrient availability, dependent on nearly intronfree DNA-protein syntax (Jeffares et al. 1998) and containing circular genomes with only one starting-point for replication. It could be considered that phages and plasmids played important roles in this natural genetic content editing because they are obligate settlers of bacteria. As intron- rich linear chromosomes are the preferred habitat for persistent retroviral infections, and because of their important role in host-genetic content (re)arrangements, the invention of bacterial circular genomes must have had an effective immune function against retroviral infections. The result was the evolution of organisms that successfully escaped the high selective pressures of the early RNA world.

References

Bapteste E, Charlebois RL, MacLeod D et al. (2005) The two tempos of nuclear pore complex evolution: highly adapting proteins in an ancient frozen structure. Genome Biol 6:R85

Birgisdottir AB, Johansen SD (2005) Reverse splicing of a mobile twin-ribozyme group I intron into the natural small subunit rRNA insertion site. Biochem Soc Transact 33:482–484

Calvin K, Li H (2008) RNA-splicing endonuclease structure and function. Cell Mol Life Sci 65:1176–1185

Chiang CC, Kennell JC, Wanner LA et al. (1994) A mitochondrial retroplasmid integrates into mitochondrial DNA by a novel mechanism involving the synthesis of a hybrid cDNA and homologous recombination. Mol Cell Biol14:6419–6432

Chiang CC, Lambowitz AM (1997) The mauriceville retroplasmid reverse transcriptase initiates cDNA synthesis De Novo at the 3' end of tRNAs. Mol Cell Biol 17:4526–4535

Daros JA, Elena SF, Flores R (2006) Viroids: an Ariadne's thread into the RNA labyrinth. EMBO Rep 7:593–598

Diener TO (1989) Circular RNAs: relics of precellular evolution? Proc Natl Acad Sci USA 86:9370–9374

Diener TO (1991) Viroids and viroid-like satellite RNAs: a phylogenetic analysis. Prog Clin Biol Res 364:243–256

Diener TO (1993) The viroid: big punch in a small package. Trends Microbiol 1:289–294

Diener TO (1995) Origin and evolution of viroids and viroid-like satellite RNAs. Virus Genes 11:119–131

Flores R, Delgado S, Gas ME et al. (2004) Viroids: the minimal non-coding RNAs with autonomous replication. FEBS Lett 567:42–48

Forterre P (2005) The two ages of the RNA world, and the transition to the DNA world: a story of viruses and cells. Biochimie 87:793–803

Forterre P, Gribaldo S, Brochier C (2005) Luca: the last universal common ancestor. Med Sci (Paris) 21:860–865

Hammer BK, Bassler BL (2007) Regulatory small RNAs circumvent the conventional quorum sensing pathway in pandemic Vibrio cholerae. Proc Natl Acad Sci USA 104:11145–11149

Itaya A, Zhong X, Bundschuh R et al. (2007) A structured viroid RNA serves as a substrate for dicer-like cleavage to produce biologically active small RNAs but is resistant to RNAInduced silencing complex-mediated degradation. J Virol 81:2980–2994

Jeffares DC, Poole AM, Penny D (1998) Relics from the RNA world. J Mol Evol 46:18–36

Majdalani N, Vanderpool CK, Gottesman S (2005) Bacterial small RNA regulators. Crit Rev Biochem Mol Biol 40:93–113

Masse E, Gottesman S (2002) A small RNA regulates the expression of genes involved in iron metabolism in Escherichia coli. Proc Natl Acad Sci USA 99:4620–4625

McClure MA (2000) The complexities of genome analysis, the retroid agent perspective. Bioinformatics 16:79–95

Obbard DJ, Gordon KHJ, Buck AH et al. (2009) The evolution of RNAi as a defence against viruses and transposable elements. Phil Trans R Soc B 364:99–115

Penny D, Poole A (1999) The nature of the last universal common ancestor. Curr Opin Genet Dev 9:672–677

Poole AK, Jeffares DD, Penny D (1998) The path from the RNA world. J Mol Evol 46:1–17

Poole A, Jeffares D, Penny D (1999) Early evolution: prokaryotes, the new kids on the block. BioEssays 21:880–889

Tu KC, Bassler BL (2007) Multiple small RNAs act additively to integrate sensory information and control quorum sensing in Vibrio harveyi. Genes Dev 21:221–233

Vetsigian K, Woese C, Goldenfeld N (2006) Collective evolution and the genetic code. Proc Natl Acad Sci USA 103:10696–10701

Villarreal LP (2005) Viruses and the Evolution of Life. American Society for Microbiology Press, Washington

Villarreal L (2009) Origin of Group Identity. Viruses, Addiction and Cooperation. Springer, New York

Wang MB, Bian XY, Wu LM et al. (2004) On the role of RNA silencing in the pathogenicity and evolution of viroids and viral satellites. Proc Natl Acad Sci USA 101:3275–3280

Wassarman KM (2002) Small RNAs in bacteria: diverse regulators of gene expression in response to environmental changes. Cell 109:141–144

Witzany G (2000) Life: The Communicative Structure. Libri Books on Demand, Norderstedt

Witzany G (2006a) Serial endosymbiotic theory (SET): the biosemiotic update. Acta Biotheor 54:103–117

Witzany G (2006b) Natural genome-editing competences of viruses. Acta Biotheor 54:235–253

Chapter 9
Viral Origins of Telomeres and Telomerases

Abstract The biocommunicative approach investigates rule-governed, sign-mediated interactions both within and among cells, tissues, organs and organisms. It also investigates genetic sequences as codes/texts that are coherent with the laws of physics and chemistry but, in addition, follow a complementary mix of combinatorial (syntactic), context-sensitive (pragmatic), content-specific (semantic) rules. In this respect, the roles of telomeres and telomerases in evolution, structure and content arrangement of genomes are of particular interest. This involves deciphering the relationships between the 'molecular syntax' of telomere repeats and their meaning, i.e. their function in the genomic content. This requires their evolutionary roots to be examined. The telomere replication process by telomerase is the most important feature here because it is processed by a very ancient competence, i.e. reverse transcriptase with a great variety of functions in most key processes of living nature.

9.1 Introduction

Upon close examination the specific characteristics of telomeres reveals certain features common to all genomes that possess telomeres: telomeres are highly conserved, non-mobile, repetitive DNA sequences. Telomeres are nucleoprotein structures that protect the ends of chromosomes from erosion, degradation, colonization, or adhering chromosome ends (Blasco 2007). They are necessary only in linear chromosomes, not in circular ones. Telomere repeats are building nodes. These nodes stabilize telomeres and *do* not consist of linear DNA. These nodes also prevent recognition as DNA damage, which would induce a DNA repair pathway. Intact nodes serve as a signal for the cell that it is fit for further replication. Telomeres are thought to be the forerunners of centromeres which probably derived from an ancestral telomere–telomere fusion (Ijdo et al. 1991). Similar to telomeres, centromeres are highly conserved, non-mobile, repetitive DNA sequences. They interact with spindle microtubules and are therefore crucial for distributing chromosomes to offspring cells (Villasante et al. 2007). They also encode small RNAs which are responsible for heterochromatin formation (Couzin 2002; Grewal and Elgin 2007).

G. Witzany, *Biocommunication and Natural Genome Editing*,
DOI 10.1007/978-90-481-3319-2_9,

Linear chromosomes of eukaryotes have the so-called end replication problem: DNA polymerases which replicate leading strands of double-stranded DNA only in the 5′ to 3′ direction are unable to replicate lagging strands, i.e. in the 3′ to 5′ direction. For leading strand replication, DNA polymerases add polynucleotides to an RNA primer. These RNA strands are later replaced by DNA. At the terminal end of the chromosome, the RNA primer cannot be replaced completely by DNA, so it cannot code for proteins or further replications. When the last RNA is added, DNA polymerase and DNA ligase transform the RNA of the primer to DNA. This process requires the presence of another DNA strand in front of the RNA primer. The end replication problem is the lack of another DNA strand in front of the last attached RNA primer. That RNA is degraded by enzymes. Thus, a section of telomeres would be lost during each replication cycle to replicate a completely lagging strand, another technique is necessary. A reverse transcriptase known as telomerase uses its integrated subunit, an inherent RNA template, to replicate the overhanging RNA primer. This allows the terminal end of the lagging strand to be fully completed without loss (Haoudi and Mason 2000).

Telomere function needs a certain length of base pairs. If this length is not available due to continued end replication problems or damage, then chromosome ends are unprotected (Du and Traktman 1996). This has prompted the suggestion that continuous telomere shortening is a main reason for cell ageing. However, recent research has documented that this is the case only in rare situations, not in general (Laun et al. 2007).

9.2 Different Molecular Syntax of Telomere Sequences

From the biosemiotic perspective it would be of interest to determine whether the telomere sequences differ between various organisms, species and kingdoms. The lack of a difference would indicate that the telomere repeat function depends on strict sequence order whereas differences would indicate that the specific function of telomere repeats is of primary importance, not the sequence order that encodes this function.

Interestingly, the molecular syntax of telomere repeats differs in those organisms in which it has been identified. This indicates that no unique molecular syntax is necessary to guarantee the function that telomere repeats have to fulfil. Rather, the same important function can be coded by different nucleic acid syntax. For instance, we find: TTAGGG in vertebrates, humans, mice, *Xenopus*, filamentous fungi, *Neurospora crassa*, the slime moulds *Physarum* and *Didymium*: TTGGGG in *Tetrahymena* and *Glaucoma*; TTGGG(T/G) in *Paramecium*; TTTTGGGG in *Oxytricha, Stylonychia* and *Euplotes*; TTAGGG(T/C) in the apicomplexan protozoan *Plasmodium*; TTTAGGG in *Arabidopsis thaliana*; TTTTAGGG in green algae *Chlamydomonas*; TTAGG in the insect *Bombyx mori*; TTAGGC in the roundworm *Ascaris lumbricoides*; TTAC(A)(C)G(1–8) in the fission yeast *Schizosaccharomyces pombe*; TGTGGGTGTGGTG (from

RNA template) in *Saccharomyces cerevisiae*; GGGGTCTGGGTGCTG in *Candida glabrata*; and GGTGTACGGATGTCTAACTTCTT in *Candida albicans*.

Telomeres act as immune functions against genomic agents with high recombination or degradation competences, i.e. viral genetic parasites, and seems to function similar to an RNAi system. RNAi protects the genome against genomic parasites, i.e. viruses, by silencing genomic transcripts of exogenous infective RNA viruses or endogenous transposons or retroposons (Fire et al. 1998; Couzin 2002; Fire 2005; Obbard et al. 2009). In addition, telomeres serve as recognition sequences, primer functions and genetic/genomic raw material for sequence generation (genome duplication, RNA template).

In *Drosophila* and some plants, telomere elongation during replication does not occur by telomerase but through recombination facilitated by the non-LTR retroposons HetA and TART (Nakamura and Cech 1998; Fajkus et al. 2005; Blasco 2007). They transport their *gag* protein into the nucleus to produce more copies of the chromosome ends (Rashkova et al. 2002). These retroposons, which fulfil the same function of telomere elongation as telomerase, are regulated by the same epigenetic regulations that govern mobile element activity, including RNAi (Savitsky et al. 2006; Slotkin and Martienssen 2007).

9.3 Telomere Replication in Most Cases by Telomerase

In most cases, except that described above, telomeres are replicated by telomerase, a reverse transcriptase. This indicates that the function of telomeres in the eukaryotic replication cycle is very ancient (Curcio and Belfort 2007). Some authors have suggested that reverse transcriptases derived from RNA-dependent RNA polymerases which themselves derived from an ancient RNA world (Boeke 2003).

Telomerase is a ribonucleoprotein enzyme that is an assembly of telomerase RNA and telomerase reverse transcriptase (Jady et al. 2004). Telomerase is clearly related to mobile elements, especially to the non-LTR retroposons (Eickbush 1999).

9.3.1 Reverse Transcriptases and Mobile Elements

Mobile elements in the genome may be transposons that integrate directly into a host genome, or retroposons that integrate via an RNA intermediate, reverse transcriptase. Copying from RNA into DNA generally involves reverse transcriptases. Mobile elements are important for genotype processing, with far-reaching consequences for phenotype expression during its various developmental stages. Recent research has demonstrated that overlapping epigenetic marking in eukaryotic cells is an important evolutionary feature to silence the expression of mobility of these mobile elements (Slotkin and Martienssen 2007). Mobile elements can silence single genes as well as larger chromosomal regions and, therefore, play an important role in the evolution of diversity. They share their competence to recombine, rearrange and insert into genomic content with other retroelements (Coffin et al. 1997). They influence neighbouring genes through alternative splicing and are active agents as enhancers and promotors or act by polyadenylation patterns (Slotkin and Martienssen 2007).

Reverse transcriptases play key roles in mobile elements like transposons and retroposons. One type of retroposon has direct repeats at its ends (LTR), others do not (non-LTRs). Interestingly, the number of retroposons increases with every transposition (transposition duplication) so that they can expand genomes: LINE-1 is 20% of the human genome (Maita et al. 2004). In contrast, transposons contain a code for the transposase protein. This enzyme identifies the terminal inverted repeats which flank mobile elements, excises them and *integrates itself instead* of those excised. The gap at the donor site is repaired in a cut-and-paste transposition or filled up with a copy of the transposon by a gap repair technique (Slotkin and Martienssen 2007). Transposons can also *integrate themselves* in phages and plasmids, and are transferred with them into other cells (Frost et al. 2005). This is evidence for a self/non-self differentiation competence.

In contrast to non-mobile telomeres and centromeres, mobile sequences such as transposons and retroposons (Volff 2006) and non-coding repetitive elements such as LTRs, SINEs and LINEs enable far-reaching DNA rearrangement and reorganization (Shapiro 2002; Sternberg 2002; Shapiro and Sternberg 2005). Together, they play a decisive role in the evolution of new genomic structures (Shabalina and Spiridonov 2004; Shapiro and Sternberg 2005; Sternberg and Shapiro 2005). Interestingly, the non-coding DNA also contains the regulations of transcription, promoter, enhancer and suppressor (Bird et al. 2006). The repetitive sequences are highly species specific and are more suitable for determining species than the coding sequences (Villarreal 2005).

This does not mean that only mobile sequences represent ancient genetic settlers. The similarity of telomeres and centromeres – non-mobile repeat elements – in descent and their relatively poor loci of inverted repeats or retroelements could indicate an ancient immune function that protects both from massive invasions by genetic parasites (Nosek et al. 2006; Villasante et al. 2007).

9.3.2 Roles of Reverse Transcriptases in Natural Genome Editing

In addition, reverse transcriptases play key roles in altering genomic structures and, therefore, in evolutionary processes facilitated by natural genome editing (Witzany 2006). Reverse transcriptases are used to generate (a) copies of mRNAs which they need for integration into a genome and (b) copies of non-mRNAs such as small nucleolar RNAs, one of the largest classes of non-coding RNAs (Zemann et al. 2006) which, like DNA copies, are SINEs. SINEs can initiate new genes which code for small RNAs with regulatory competences on existing genes.

One further key feature of reverse transcriptases is that they are a primer for retroposons such as LTRs (copia, gypsy, Ty1, IAPs, HERVs). Non-LTRs (Het-A/TART,

SINEs, LINEs) act like telomerases in several arthropods and plants. Moreover, reverse transcriptases are encoded and used by open reading frames (ORF), ORF1 (an RNA-binding and shuttling protein), ORF2 (endonuclease, reverse transcriptase activities), as well as ALUs (manipulation of LINE-1 function for mobilization), group II self-splicing introns and snoRNAs (type 1–3 retroposons), all of which act as important regulatory functions (Yang et al. 1999; Batzer and Deininger 2002; Tomlinson et al. 2006; Weber 2006; Matera et al. 2007).

Reverse transcriptases are also found in retroviruses of mammals and birds, in the hepadnavirus of mammals and birds, and the caulimovirus of plants, in LTR retroposons of animals, plants, fungi and protozoa, in non-LTR retroposons of animals, plants, fungi and in protozoa, group II introns of bacteria, fungi, plant mitochondria, chloroplasts and plastids, in mitochondrial plasmids of Neurospora mitochondria, and in multiple single-stranded DNAs (Villarreal 2005). Although many researches believe that non-LTRs evolved before retroviruses, there is recent evidence that viruses are their ancestors (Villarreal 2009).

RNA-dependent RNA polymerases together with reverse transcriptases replicate positive-strand RNA viruses, double-stranded RNA viruses, negative-strand RNA viruses and retroviruses (Leipe et al. 1999; Koonin 2006; Koonin et al. 2006). The RNA-dependent RNA polymerases are initiated by two or more complementary microRNA sites (Slotkin and Martienssen 2007) that could indicate an addiction module because of its counterpart regulation. RNA-dependent RNA polymerases are involved in the coupling of heterochromatin for the production of siRNAs (Sugiyama et al. 2005). The RNAi system is competent in post-transcriptional gene silencing and is, therefore, a crucial instrument in keeping the balance between the need for expression and the need for silencing (Grewal and Elgin 2007) siRNAs therefore act similar to endogenously encoded microRNAs (Doench et al. 2003).

Many organisms have ORFs that code for proteins with sequences very similar to retroviral reverse transcriptases (Xiong and Eickbush 1990; Mesnard and Lebeurier 1991). RNA-dependent DNA polymerase (reverse transcriptase) has relations to RNA-dependent RNA polymerase. Rooting these lines of descent in RNA-dependent RNA polymerases yields two groups: (i) group 1 contains LTR retroposons, RNA viruses, DNA viruses; (ii) group 2 contains non-LTR retroposons, bacterial and other organelle parts (Nakamura and Cech 1998).

The telomerase function is cell cycle regulated. It functions exclusively if its suppression is deleted. Once the telomerase function in telomere replication is fulfilled, a signal initiates its suppression again. A disturbed signalling process may lead to uncontrolled cell replication. Telomerase has to be transported to telomere repeats for its elongation during the S phase of the cell cycle. The delivery agents are Cajal bodies – small nucleolus-like organelles competent in (i) splicing, (ii) ribosome production and (iii) transcription (Platani et al. 2002; Jady et al. 2004). They are located in the periphery of nucleoli (Darzacq et al. 2002; Matera 2006). Cajal bodies move throughout the area of the nucleus and, for certain properties, they fuse with other Cajal bodies or associate with nucleoli (Tomlinson et al. 2006, Kiss et al. 2001).

Telomerase trafficking is restricted to the S phase of the cell cycle, which avoids telomerase activity at non-telomeric sites of the chromosomes. (Tomlinson et al. 2006).

9.4 Telomeres are Characteristics of Eukarya

Because telomere repeats and telomerases are key features of eukaryotes, and not of prokaryotes, it may be concluded that eukaryotic telomeres and telomerases are interconnected with the evolution of the eukaryotic cells. Deciphering the evolutionary roots of telomeres and telomerases necessitates the main differences between eukaryotes and prokaryotes to be examined. The evolutionary agents of the eukaryotic nucleus may even point to the roots of telomeres and telomerases.

First, eukaryotic genomes share a great variety of repeat elements with higher-order regulatory functions. In contrast to prokaryotes, eukaryotic replication proteins have very different amino acid sequence compositions. In addition, eukaryotes share the control of DNA packaging and replication, whereas prokaryotes do not have chromatin proteins such as histones (Villarreal 2005).

The eukaryotic DNA replication starts in numerous (thousands) sites and is regulated by a complex cell cycle regulatory system. Eukaryotic replication control proteins do not resemble prokaryotic ones. A further difference between eukaryotes and prokaryotes is that daughter cells segregate by attachment to a microtubule system (spindles), not by attachment at the membrane. The highly conserved mitotic spindle system is not found in any prokaryote (Cottingham and Hoyt 1997).

Also, the eukaryotic nucleus possesses three classes of DNA-dependent RNA polymerases that do not resemble the polymerases of any prokaryote. A further crucial difference is that in eukaryotes the products of RNA polymerases must undergo post-transcriptional modifications (splicing) before they can function in the cytoplasm as mRNA, tRNA or rRNA. No prokaryote exhibits splicing of pre-mRNAs. To prevent mistranslation of mRNA or unspliced tRNA, the nucleus has to separate transcription/processing of mRNA from the cytoplasm transport of processed RNAs. This requires a nuclear membrane to segregate transcription, mRNA processing, transport and translation in the cytoplasm (Vale 2003). The nuclear membrane is distinct from the cell membrane and is dissolved after the S phase, but is restored at late anaphase/telophase. All complex modifications of mRNA and nuclear RNA seem to be acquired during the evolution of the eukaryotic nucleus; they are highly conserved in eukaryotes but absent in prokaryotes.

Only a very few prokaryotic genomes share some of the above-mentioned features. In the case of the spirochetes *Borrelia*, the genomes possess three types of telomeres, segmented genomes of linear and circular plasmids and extensive DNA rearrangements (Chaconas 2005; Tourand et al. 2006). This could indicate intensive infection by competing genetic parasites which are in balance as ‘addiction modules’ (see below) in a persistent status. This does not harm the host but is harmful to those organisms (even close relatives) that lack these persistent inhabitants.

9.5 Agents of Natural Genome Editing

Recent research shows extensive dynamic DNA remodelling by small RNAs and micro-RNAs, which are competent in a great variety of DNA arrangements, rearrangements and recombinations (Shapiro 2002; Vaughn and Martienssen 2005; Mattick 2001, 2006). Some authors refer to agents of genomic creativity (Ryan 2006), mobile or regulatory elements (Eickbush 1999; Brosius 2003) or entities (Daubin and Ochman 2004), while others refer to transposable elements (Slotkin and Martienssen 2007), non-coding RNA populations (Mattick 2007) and still others to mobile DNA species or genetic parasites (Nakamura and Cech 1998; Villarreal 2005). Together, these agents enable complex organisms to integrate several temporal steps and a great variety of coordinated signalling processes in eukaryotic cell replication, fix them in a conserved DNA storage medium and, if necessary, resolve conservation, change, rearrange or newly construct the whole genomic content and sequence order (Shapiro 2006).

The DNA information storage medium is and has to be edited. I predict a future discussion on how to refer to these editing agents, for example as interactions of more or less chemical molecules or as 'non-random genetic change operators' (Shapiro 2007, personal communication)

From a biocommunicative perspective – which investigates combinatorial (syntactic), content-specific (semantic) and contextual (pragmatic) rules of natural genome editing and genetic text processing – it is important to note that there can be no editing without a subject that edits, i.e. an editor or a swarm of editors (Vetsigian et al. 2006). For example, the spliceosome works as an integrated network of several small nuclear RNAs and their associated proteins on the primary RNA-transcript into the pre-mRNA (Vaughn and Martienssen 2005).

Life could not function without the key agents of DNA replication, namely mRNA, tRNA and rRNA. Not only rRNA, but also tRNA and the processing of the primary transcript into the pre-mRNA and the mature mRNA, are clearly descended from retroelements (Maizels and Weiner 1999; Maizels et al. 1999; Flavell 1995; Eickbush and Eickbush 2007)

It is now possible to appreciate how sophisticatedly the competent, subject-like operators act in the case of endogenous retroviruses, which reached a persistent and non-lytic lifestyle. We also know that all related retroelements share a common genome-editing competence like transposable 'elements'. Nonetheless, it remains difficult to reconstruct how all these DNA-encoded RNA agents reached persistent status in hundreds, thousands and tens of thousands of elements. We only know that they act in a precisely coordinated manner which would be impossible without competent signalling. This includes a strict competence for self/non-self identification, which is a major asset of RNAs in general and of small nucleolar RNAs in particular (Filipowicz 2000).

Persistent endogenous agents competent in both natural genetic engineering and natural genome editing apparently prefer a special kind of habitat characterized as non-coding DNA sectors. They use a syntax mainly consisting of repeats. They colonized analogous DNA genomes by inserting their sites between coding elements;

then they use these coding elements for different needs. This developed to the point that, in the human genome, only 3% of coding regions remained. The remaining 97% serves as a habitat for persistent viral operators that orchestrate a highly sophisticated division of labour. From these genomic locations they can actively regulate close coding-sequences. Of special interest is the highly sophisticated production of mRNA with its cut-and-paste process in which non-coding elements, i.e. introns, are spliced out; the remaining exons which code for proteins are combined into a coherent protein-coding content ready for translation.

Persistent endogenous agents in some cases are intact as full length agents being important actors at certain replication stages as demonstrated in placentation of mammals. In most cases former intact colonizing viral agents are split into parts such as gag, pol or env or transposons and retroposons such as the whole range of long terminal repeats (LTR's), SINEs or LINEs and ALUs which from the viral perspective now are 'defectives', i.e. non-functional in the former viral competence. But from the host perspective these parts now play important roles in gene-regulation, genome-elongation or genetic new content arrangements. Therefore these 'defectives' now represent 'effectives' for the host and changed both the viral and the host identity.

As opposed to persistent endogenous agents of natural genome editing in eukaryotes, we find persistent exogenous agents in prokaryotes that are competent in natural genome editing in the prokaryotic gene pool. This process has long been visualized as horizontal gene transfer and is now recognized as occurring by plasmids, phages and transposons, all with viral ancestors (Frost et al. 2005).

It is difficult to perceive mere molecules or molecule buildings as being 'competent' to process the sophisticated DNA language. It is less difficult to think of viruses being these subject-like agents.

9.6 Superficial and Deep Grammar in Eukaryotic Genome Content

Higher-order regulations which are performed by agents inherent in non-coding RNAs and in most repeat elements such as subtelomeric repeats and all the other retroelements have a similar relationship to protein-coding sequences as operators competent in using a (1) deep grammar with which they determine (2) the superficial grammar of sequence content. Through these two different levels it is possible to determine the protein-coding data sets, according to different needs, into 'multiple protein meanings' (Ast 2005). Eukaryotic genome evolution involved the step from a continous coding sequence order to an interrupted sequence order. Interestingly, the former is characteristic for circular prokaryotic genomes, the latter for linear genomes.

From the biocommunicative perspective this symbiogenetically induced innovation of multiple-invaded coding data sets by retroelements opened up the possibility of using protein-coding data sets according to various types of higher-order regulation. The protein-coding data sets are the structural vocabulary, the non-protein-coding 'underworld' (Mattick 2006) of RNAs is the text-editing operators. This involved a massive invasion by non-coding introns (viruses) into the genomic habitats of protein-coding data sets (Rogozin et al. 2005; Mattick 2007). Thus, the molecular syntax of protein-coding data sets could be used for different requirements in different contexts (pragmatics) to serve for different genetic content arrangements (semantics). This could explain:

- that in evolutionary history certain genotypes from one species are transferred and integrated into the genomic content of other species to yield a new role or a new phenotypic feature in another context. This occurred with telomeres in the linear chromosomes of ancient double-stranded DNA viruses (poxvirus, vaccinia virus, archaeal phages: AFV-1, SIRV-1, TTV 1–4), where they had other functions than in the eukaryotic genomic content (Villarreal 2005);
- the close coherence of protein-coding data sets between humans and chimpanzees (99%), keeping in mind that the percentage of protein coding in humans and chimpanzees is only 3%, whereas the percentage of non-coding DNA with higher-order regulatory functions is 97%, which determines different expression patterns (Witzany 2006);
- that specific cellular functions are encoded in a weakly conserved manner at the sequence level, in contrast to their preserved domains, for example the genes of nuclear pores (Bapteste et al. 2005);
- that telomeres themselves are not typical sites for colonization events, in contrast to sites very close to these telomeres. This is similar to the phylogenetically related centromeres. Because telomeres and centromeres themselves are relatively free of inverted repeats or retroelements, this could indicate an ancient immune (RNAi) function that protects both from massive invasions by genetic parasites.

9.7 Conclusion

Whereas telomeres protect terminal ends of linear chromosomes, telomerases identify natural chromosome ends, which differ from broken DNA. Although telomeres play a crucial role in the linear chromosome organization of eukaryotic cells, their molecular syntax most probably descended from an ancient retroviral competence. This indicates an early retroviral colonization of large double-stranded DNA viruses, which are putative ancestors of the eukaryotic nucleus. This contribution demonstrates an advantage of the biocommunicative approach towards our evolutionary understanding of telomeres, telomerases, other reverse transcriptases and mobile elements. Their role in genetic/genomic content organization and maintenance is no longer viewed as an object of randomly derived alterations (mutations) but as a highly sophisticated hierarchy of regulatory networks organized and coordinated by natural genome-editing competences of viruses.

The acquisition of telomere repeats in eukaryotes was a key event in eukaryotic nucleus evolution (Eickbush 1997). The eukaryotic nucleus most probably evolved from a large DNA virus. The changing structure of the eukaryotic genome, however, with its coding and non-coding sections and its typical repetitive (higher-order regulatory) elements, indicates high rates of persistent, non-lytic viral infections. In contrast to most of these mobile, higher-order regulatory agents, telomere repeats (as well as centromeres) attained a non-mobile status. This indicates an ancient stable conservation (Nosek et al. 2006).

Telomerase, from the biocommunicative perspective, is a natural genetic engineering tool with different functions in different contexts. Whereas in the RNA virus life cycle reverse transcriptase is used for replication functions, it serves as an acquired tool for complete replication of chromosomal ends in linear eukaryotic genomes. In eukaryotes, telomerases and other reverse transcriptases act as 'effective'endogenous viral competences.

In symbiogenetic infection events, the eukaryotic host acquired a higher-order regulated genomic syntax. This is the precondition for multiple protein meanings from the same genetic data set through post-transcriptional modifications such as alternative splicing pathways. The transformation of the continuous (prokaryotic) molecular syntax into a eukaryotic molecular syntax invaded by a great diversity of natural genome-editing agents is, therefore, a major step in the evolution of multicellular complexity. Highly conserved protein-coding sequences are object to a great variety of 'effective' agents being competent in the generation of modular genetic content arrangements for different needs in host cell replication.

References

Ast G (2005) The alternative genome. Sci Am 292:58–65

Bapteste E, Charlebois RL, MacLeod D et al. (2005) The two tempos of nuclear pore complex evolution: highly adapting proteins in an ancient frozen structure. Genome Biol 6:R85

Batzer MA, Deininger DL (2002) ALU repeats and human genomic diversity. Nat Rev Genet 3:370–380

Bird CP, Stranger BE, Dermitzakis ET (2006) Functional variation and evolution of non coding DNA. Curr Opin Genetics Dev 16:559–564

Blasco M (2007) The epigenetic regulation of mammalian telomeres. Nat Rev 8:299–309

Boeke JD (2003) The unusual phylogenetic distribution of retrotransposons: a hypothesis. Genome Res 13:1975–1983

Brosius J (2003) The contribution of RNAs and retroposition to evolutionary novelties. Genetica 118:99–115

Chaconas G (2005) Hairpin telomeres and genome plasticity in Borrelia: all mixed up in the end. Mol Microbiol 58:625–635

Coffin JM, Hughes AH, Varmus HE (1997) Retroviruses. Cold Spring Harbor Laboratory Press, New York

Cottingham FR, Hoyt MA (1997) Mitotic spindle positioning in saccharomyces cerevisiae is accomplished by antagonistically acting microtubule motor proteins. J Cell Biol 138:1041–1053

Couzin J (2002) Small RNAs make big splash. Science 298:2296–2297

Curcio MJ, Belfort M (2007) The beginning of the end: links between ancient retroelements and modern telomerases. Proc Natl Acad Sci USA 104:9107–9108

Darzacq X, Jady BE, Verheggen C et al. (2002) Cajal body-specific small nuclear RNAs: a novel class of 2'-O-methylation and pseudouridylation guide RNAs. EMBO J 21:2746–2756
Daubin V, Ochman H (2004) Start-up entities in the origin of new genes. Curr Opin Genetics Dev 14:616–619
Doench JG, Petersen CP, Sharp PA (2003) siRNAs can function as miRNAs. Genes Dev 17: 438–442
Du S, Traktman P (1996) Vaccinia virus DNA replication: two hundred base pairs of telomeric sequence confer optimal replication efficiency on minichromosome templates. Proc Natl Acad Sci USA 93:9693–9698
Eickbush TH (1997) Telomerase and retrotransposons: which came first? Science 277:911–912
Eickbush T (1999). Mobile introns: retrohoming by complete reverse splicing. Curr Biol 9: 11–14
Eickbush TH, Eickbush DG (2007) Finely orchestrated movements: evolution of the ribosomal RNA genes. Genetics 175:477–485
Fajkus J, Sykorova E, Leitch AR (2005) Telomeres in evolution and evolution of telomeres. Chromosome Res 13:469–479
Filipowicz W (2000) Imprinted expression of small nucleolar RNAs in brain: time for RNomics. Proc Natl Acad Sci USA 97:14035–14037
Fire A (2005) Nucleic acid structure and intracellular immunity: some recent ideas from the world of RNAi. Q Rev Biophys 38:303–309
Fire A, Xu S, Montgomery MK et al. (1998) Potent and specific genetic interference by double-stranded RNA in Caenorhabditis elegans. Nature 391:806–811
Flavell AJ (1995) Retroelements, reverse transcriptase and evolution. Comp Biochem Physiol B 110:3–15
Frost LS, Laplae R, Summers AO et al. (2005) Mobile genetic elements: the agents of open source evolution. Nat Rev Microbiol 3:722–732
Grewal SIS, Elgin SCR (2007) Transcription and RNA interference in the formation of heterochromatin. Nature 447:399–406
Haoudi A, Mason JM (2000) Reverse transcriptase can stabilize or destabilize the genome. Genome 43:949–956
Ijdo JW, Baldini A, Ward DC et al. (1991) Origin of human chromosome 2: an ancestral telomere-telomere fusion. Proc Natl Acad Sci USA 88:9051–9055
Jady BE, Bertrand E, Kiss T (2004) Human telomerase RNA and box H/ACA scaRNAs share a common Cajal body-specific localization signal. J Cell Biol 164:647–652
Kiss AM, Jady BE, Darzaq X et al. (2001) A Cajal body specific pseudouridylation guide RNA is composed of two box H/ACA snoRNA-like domains. Nucleic Acids Res 30:4643–4649
Koonin EV (2006) Temporal order of evolution of DNA replication system inferred by comparison of cellular and viral DNA polymerases. Biology Direct doi:10.1186/1745-6150-1–39
Koonin EV, Senkevich TG, Dolja VV (2006) The ancient Virus World and evolution of cells. Biol Direct 1:29
Laun P, Bruschi CV, Dickinson JR et al. (2007) Yeast mother cell-specific ageing, genetic (in)stability, and the somatic mutation theory of ageing. Nucleic Acids Res doi:10.1093/nar/gkm919, 1–14.
Leipe DD, Aravind L, Koonin EV (1999) Did DNA replication evolve twice independently? Nucleic Acids Res 27:3389–3401
Maita N, Anzai T, Aoyagi H et al. (2004) Crystal structure of the endonuclease domain encoded by the telomere-specific long interspersed nuclear element, TRAS1. J Biol Chem 279:41067–41076
Maizels A, Weiner AM (1999) The genomic tag hypothesis: modern viruses as molecular fossils of ancient strategies for genomic replication. Biol Bull 196:327–330
Maizels N, Weiner AM, Yue D et al. (1999) New evidence for the genomic tag hypothesis: archaeal CCA-adding enzymes and tRNA substrates. Biol Bull 196:331–334
Matera AG (2006) Drosophila Cajal bodies: accessories not included. J Cell Biol 172:791–793

Matera AG, Terns RM, Terns MP (2007) Non-coding RNAs: lessons from the small nuclear and small nucleolar RNAs. Nat Rev Mol Cell Biol 8:209–220
Mattick JS (2001) Non-coding RNAs: the architects of eukaryotic complexity. EMBO Rep 2: 986–991
Mattick JS (2006) The underworld of RNA. Nat Genet 38:393
Mattick JS (2007) A new paradigm for developmental biology. J Exp Biol 210:1526–1547
Mesnard JM, Lebeurier G (1991) How do viral reverse transcriptases recognize their RNA genome? FEBS Lett 287:1–4
Nakamura TM, Cech TR (1998) Reversing time: origin of telomerase. Cell 92:587–590
Nosek J, Kosa P, Tomaska L (2006) On the origin of telomeres: a glimpse at the pre-telomerase world. Bioessays 28:182–190
Obbard DJ, Gordon KHJ, Buck AH et al. (2009) The evolution of ENAi as a defence against viruses and transposable elements. Phil Trans R Soc B 364:99–115
Platani M, Goldberg I, Lamond AI et al. (2002) Cajal body dynamics and association with chromatin are ATP dependent. Nat Cell Biol 4:502–508
Rashkova S, Karam SE, Kellum R et al. (2002) Gag proteins of the two drosophila telomeric retrotransposons are targeted to chromosome ends. J Cell Biol 159:397–402
Rogozin IB, Sverdlov AV, Babenko VN et al. (2005) Analysis of evolution of exon-intron structure of eukaryotic genes. Brief Bioinform 6:118–134
Ryan FP (2006) Genomic creativity and natural selection: a modern synthesis. Biol J Linn Soc 88:655–672
Savitsky M, Kwon D, Georgiev P, Kalmykova A, Gvozdev V (2006) Telomere elongation is under the control of the RNAi-based mechanism in the Drosophila germline. Genes Dev 20:345–354
Shabalina SA, Spiridonov NA (2004) The mammalian transcriptome and the function of non-coding DNA sequences. Genome Biol 5:105e
Shapiro JA (2002) Genome organization and reorganization in evolution. In: Van Speybroeck L, Van de Vijver G, Waele DD (eds) From Epigenesis to Epigenetics. The genome in context. Ann NY Acad Sci 981:111–134
Shapiro JA (2006) Genome informatics: the role of DNA in cellular computations. Biol Theor 1:288–301
Shapiro JA, Sternberg R (2005) Why repetitive DNA is essential to genome function. Biol Rev 80:1–24
Slotkin RK, Martienssen R (2007). Transposable elements and the epigenetic regulation of the genome. Nat Rev Genet 8:272–285
Sternberg R (2002) On the roles of repetitive DNA elements in the context of a unified genomic-epigenetic system. In: Van Speybroeck L, Van de Vijver G, Waele DD (eds) From Epigenesis to Epigenetics. The genome in context. Ann NY Acad Sci 981:154–188
Sternberg R, Shapiro JA (2005) How repeated retroelements format genome function. Cytogenet Genome Res 110:108–116
Sugiyama T, Cam H, Verdel A et al. (2005) RNA-dependent RNA polymerase is an essential component of a self-inforcing loop coupling heterochromatin assembly to siRNA production. Proc Natl Acad Sci USA 102:151–157
Tomlinson RL, Ziegler TD, Supakorndej T et al. (2006) Cell cycle-regulated trafficking of human telomerase to telomeres. Mol Biol Cell 17:955–965
Tourand Y, Bankhead T, Wilson SL et al. (2006) Differential telomere processing by borrelia telomere resolvases in vitro but not in vivo. J Bacteriol 188:7378–7386
Vale R (2003) The molecular motor toolbox for intracellular transport. Cell 112:467–480
Vaughn MW, Martienssen R (2005) It's a small RNA world, after all. Science 309:1525–1526
Vetsigian K, Woese C, Goldenfeld N (2006) Collective evolution and the genetic code. Proc Natl Acad Sci USA 103:10696–10701
Villarreal LP (2005) Viruses and the Evolution of Life. ASM Press, Washington
Villarreal LP (2009) Origin of Group Identity: Viruses, Addiction and Cooperation. Springer, New York

Villasante A, Abad JP, Mendez-Lago M (2007) Centromeres were derived from telomeres during the evolution of the eukaryotic chromosome. Proc Natl Acad Sci USA 104:10542–10547

Volff JN (2006) Turning junk into gold: domestication of transposable elements and the creation of new genes in eukaryotes. Bioessays 28:913–922

Weber MJ (2006) Mammalian small nucleolar RNAs are mobile genetic elements. PLoS Genetics doi:10.1371/journal.pgen.0020205

Witzany G (2006) Natural genome-editing competences of viruses. Acta Biotheor 54:235–253

Xiong Y, Eickbush TH (1990) Origin and evolution of retroelements based upon their reverse transcriptase sequences. EMBO J 9:3353–3362

Yang J, Malik HS, Eickbush TH (1999) Identification of the endonuclease domain encoded by R2 and other site-specific, non-long terminal repeat retrotransposable elements. Proc Natl Acad Sci USA 96:7847–7852

Zemann A, Beckke A, Kiefmann M et al. (2006) Evolution of small nucleolar RNAs in nematodes. Nucleic Acids Res 34:2676–2685

Chapter 10
Real Life-World of Noncoding RNA-Species

Abstract In the last decade it was found that the number of genes of some nematodes and humans was similar but their regulation was completely different. Today we know that the higher-order regulation of protein-coding datasets depends on complex interconnected networks of a great variety of non-coding RNAs that are read and transcribed in the developmental and growth processes of every cell within multicellular organisms. The evolutionary origins of these non-coding RNAs are not randomly-derived mixtures of nucleotide acids but formerly intact viral agents which infected all cellular host genomes in a non-lytic but persistent way. Although some of these viral agents still fulfill vital functions, e.g., endogenous retroviruses which are active in placentation of mammals, in most cases they split up ('defectives') into several functional parts which now serve as 'effectives', i.e. symbiogenetic integrated functional tools for cellular needs of host organisms.

10.1 Introduction

As proposed in an article by Vetsigian et al. (2006), DNA in contrast to RNA serves as an 'evolutionary protocol' for evolutionary novelties and selected properties which indicates DNA as a stable information storage medium. A wide variety of small RNAs regulate key cellular processes of replication as well as genetic arrangements, rearrangements, recombination and repair and even inventions. In most cases they act after being transcribed from the stable DNA storage medium into a pre-transcriptional or transcriptional modus, with the advantage of being active prior to all translated proteins. Elements such as microRNAs, small nuclear and small nucleolar RNAs, tRNA and rRNA as well as the assemblies of the spliceosome and ribosome are vital to all life processes. In this respect I remember a quote from the Austrian philosopher Ludwig Wittgenstein: 'The meaning of a word is its use within a language' and 'to understand a sentence means to understand a language. To understand a language means to be the master of a technique'.

G. Witzany, *Biocommunication and Natural Genome Editing*,
DOI 10.1007/978-90-481-3319-2_10,

What are the 'masters' of the technique used to edit nucleotide sequences of the genetic text according to combinatorial (syntactic), context-sensitive (pragmatic) and content-specific (semantic) rules, according to Charles Morris the obligate and non-reducable levels of rules that are inherent to any kind of language or language-like codes?

10.2 Genetic Text-Sequences Function Similar to any Natural Language

Several decades it was considered that the genetic content arrangements of DNA sequences are evolutionary results of random mutations and their selection. Then it was noticed that DNA regions which code for proteins are decreasing with organismic complexity (in humans 1,5%) and former 'junk DNA' plays increasing important roles in gene regulation. Additionally, it was assumed that this 'junk DNA' is a kind of selfish genetic elements, with an inherent tendency to replicate themselves, in contrast with its helpful and symbiotic functions within host genomes.

For many decades it was common practice to speak about the 'genetic code' with its inherent language-like features. Long before, and even after, Manfred Eigen's suggestion that the nucleic acid sequences are comparable to and function like any real language coherent with a (molecular) syntax, linguistic and communicative vocabulary was commonly used in genetics, cell biology and molecular biology: genetic code, code without commas, misreading of the genetic code, coding, genetic storage medium DNA, genetic information, genetic alphabet, genetic expression, messenger RNA, cell to cell communication, immune response, transcription, translation, nucleic acid language, amino acid language, recognition sequences, recognition sites, protein coding sequences, repeat sequences, etc.

In contrast with the evolutionary paradigm of random assemblies of nucleic acids which constitute the genetic text we do not know any real-life languages or codes which emerged as a randomly derived mixture of the characters of an alphabet.

If Manfred Eigen's suggestion is still valid, and the description of nucleic acid sequence order in terms of linguistics (molecular syntax) makes sense also for the future, we should look at the current scientific knowledge of 'language' and 'communication'.

Every language is based on signs, whether they are signals or symbols. In humans and other animals they are transported auditively, visually or tactilely. In non-human living beings they are transported by small molecules in crystallised, fluid, gaseous form. Additionally these signs can be combined coherently with combinatorial rules (syntax). Signs are not generated and used by themselves, but in real-life languages by living beings. These sign-generating and sign-using agents live in vivo in continued changing interactions and situational and environmental circumstances. This is the context (pragmatics) in which a living being is interwoven. This context determines the meaning (semantics) of the signs in messages which are used to communicate and to coordinate single as well as group behaviour.

Therefore we can understand that the same sentence, or the same syntactic sequence order, of any language or code can have different and in extreme cases opposite meanings and therefore transport different messages. *The important consequence of this fact is that it is not possible to extract the meaning of an informational content solely out of the syntactic structure but someone has to identify the context within which the living being uses this syntactic structure.*

The primary agents are not the sequences of signs, not the rules which determine sequence orders, but the living agents. Without living agents there are no signs, no semiotic rules, no signalling and no communication. Paradoxically, without signs, semiotic rules, signalling and communication, no living agents could coordinate growth and development.

If we assume the genetic code to function language-like, knowing that no language which has been observed functions by itself, then we have to postulate living agents which are competent to use signs coherent with syntactic, pragmatic and semantic rules. Adapted to the genetic code, this means both that there must be living agents competent in generation and integration of meaningful nucleotide sequences, and meaningful nucleotide sequences are not a randomly derived mixture of nucleotides.

Natural genome editing from a biocommunicative perspective therefore means competent agent-driven generation and integration of meaningful nucleotide sequences into pre-existing genomic content arrangements and the ability to (re-)combine and (re-)regulate them according to the context-dependent (i.e. adaptational) purposes of the host organism.

10.3 Cellular DNA Nucleotide Sequences as Viral Life Habitat

As we know today all cellular life is colonized by exogenous and/or endogenous viruses in a non-lytic but persistent lifestyle. A persistent lifestyle in cellular life-forms most often seems to derive from an equilibrium status reached by at least two competing genetic settlers and the immune function of the host which keeps them in balance. Persistent settlement of host genomes means that if we postulate agent-driven genetic text editing then we have to look at their in vivo life-strategies to understand their habits and the situational contexts that determine their content arrangements. Then we can reconstruct nucleic acid sequences that function as a code, not as a statistically random-like mixture of nucleotides but as informational content in a syntactic order that is coherent with the whole sequence space generated by agents that are linguistically competent in nucleic acid language, i.e. the genetic code. As in every language each character, word and sentence together with starts, stops, commas and spaces in-between has content and a text-formatting function and is generated by competent agents.

If we imagine that humans and one of the simplest animals, *C. elegans*, share a nearly equal number of genes (ca. 20,000) it becomes obvious that the elements that create the enormous diversity are not the protein-coding genes but their higher order regulatory network which is processed by the mobile genetic elements such as transposons and retroposons and the non-coding RNAs (Claverie 2005). Juergen Brosius created the appropriate image of 'genes floating in a sea of retroposons'. If we consider the important role of the highly structured and ordered regulatory network of non-coding RNAs as not being randomly derived, one of the most favorable models with explanatory power is the virus-first thesis. This means that the evolution of the non-coding RNA-world is the result of persistent viral life-strategies.

The whole range of mobile genetic agents that are competent to edit the genetic code/nucleic acid language, not only edit but also regulate key cellular processes of replication as well as transcription, translation, recombination, repair and even inventions via a wide variety of small RNAs. In this respect, DNA is not only an information-storing archive but a life-habitat for linguistically-competent RNA agents, most of them seemingly of viral or subviral descent. To understand their natural genome-editing-competence we have to look not only at their *linguistic* competence in editing and regulating correct nucleotide sequences but at their *communicative* competence too, i.e. how they interact with each other, how they compete within host organisms, how they symbiotically interact with host organisms to ward off competing parasites, and what life-strategies they share. Persistent infection-lifestyles that do not harm hosts and symbiotic, cooperating viral swarms may be more successful in evolutionary terms for integrating advantageous phenotypes into host organisms than 'selfish' agents.

10.3.1 The Persistence of the Eukaryotic Nucleus

The eukaryotic cell most probably evolved by a symbiogenetic integration event of former free-living bacteria. This integration, however, cannot explain the progenitor of the eukaryotic nucleus because its key features could not have derived from prokaryotes (Bell 2001, 2006). The eukaryotic nucleus has numerous key features, proteins and RNAs that are *not found in any prokaryote*. Interestingly, these key features are present in certain prokaryote viruses (Villarreal 2005; Forterre 2006a; b). These viruses use linear chromosomes, telomere repeats, multiple membranes, histone-packaged chromosomes with marking effect for self/non-self identification, and nuclear pores.

No single virus encompasses all of these key features, but every key feature of the eukaryotic nucleus is present in some large dsDNA viruses. This requires consideration of a process in which different viral competences were integrated into a single dsDNA virus that was the progenitor of the eukaryotic nucleus. Alternatively, a large dsDNA virus functioned as a simple eukaryotic nucleus and later integrated other viral competences. On examination of the key features of several candidates for this integration, the focus is primarily on prokaryotic, eukaryotic and archaeal phages.

Prokaryotic phages such as cyanophages have double-stranded DNA, DNA polymerases and RNA polymerases similar to eukaryotes. Eubacterial phages possess linear double-stranded DNA, telomeres, DNA polymerases, RNA polymerases, chromatin and internal membranes. Archaeal phages with linear double-stranded DNA have telomere repeats similar to eukaryotes. They also possess chromatin and an internal lipid tendency to non-lytic, persistent (and often mixed) infections (Villarreal 2005).

Other DNA viruses share similar features that are characteristic of the eukaryotic nucleus but are not found in prokaryotes. An example is the vaccinia virus (poxvirus) (Takemura 2001). These viruses have a membrane-bound segregation of transcription and translation, multiple membranes, and their DNA synthesis combines membrane loss and a cell cycle-dependent restoration as well as an actin/tubulin-bound transport system (Villarreal 2005; Van Lent and Schmitt-Keichinger 2006) and, interestingly, nuclear pores (Bapteste et al. 2005). Cytoplasmic DNA viruses (African swine fever virus) have chromatin and linear chromosomes with telomeres. PhycoDNA viruses have mRNA capping, introns and diverse DNA replication proteins. TTV (1–4) have linear double-stranded DNA genomes with a molecular basis for the evolution of eukaryotic chromatin; they also have capsids which integrate internal and external lipid proteins (Villarreal 2005).

In addition, all these viruses have the capability for self and non-self identification. All viruses mark their genomes, RNAs and proteins by different kinds of chemical modifications, e.g. methylation. This marking allows the differentiation between self and non-self. Non-self may be other viruses, the host genome or host-related transcripts (Villarreal 2005).

10.4 Viral Agents as Genetic Editors

The evolution of the eukaryotic nucleus from this perspective seems to be the result of the natural genome-editing competences of viruses (Witzany 2007a). Recent research in microbiology, based on comparative genomics and phylogenetic analyses, has demonstrated that life must be viewed from the perspective of the crucial role played by viruses (Villarreal 2005; Forterre 2001, 2002, 2005; Koonin et al. 2006; Tran et al. 2004).

This contradicts former concepts which focused on viruses in the framework of (i) escape theories, i.e. viruses are intact or deformed genetic parasites which escaped from cellular life, or considered that viruses (ii) evolved from cellular ancestors or (iii) that they are not living beings because they cannot live without cellular life. From these perspectives, viruses could not play crucial roles in the evolution of cellular life. Interestingly, phylogenetic analyses do not support the former concept of RNA- and DNA-viruses descending from cellular life. These analyses also show that DNA viruses and RNA viruses most probably did not have a common ancestor but evolved independently. Viruses probably have to be placed at the very beginning of life, long before cellular life evolved (Villarreal 2005).

10.4.1 Persistent Viral Life Strategies Change Genetic Host-Identities

In contrast to acute viruses that exhibit lytic action that induce disease and even death, a persistent lifestyle implies compatible interactions with the host, either by being integrated into the host genome or within its cell plasma (Gorinsek et al. 2004). The result is non-destructive symbiosis during most life stages of the host. The persistent lifestyle allows the virus to transmit complex viral phenotypes to the host organism. This process, which changes both the genetic identity of the host and the identity of its persistent settler, enables the host to broaden its evolutionary, adaptational potential and may promote the formation of new species (Villarreal 2005).

The persistent lifestyle of viruses is typically tissue specific, i.e. host tissues are colonized by different non-lytic viruses which integrate themselves into the host cytoplasm, e.g. as plasmids or into the host genomes, and co-evolve with them. A common habit of persistent viral settlers is that during host cell replication they function in a tissue-specific, replication-cycle dependent manner. Interestingly, micro-RNAs in eukaryotic cells have similar tissue-specific or developmental expression patterns (Mattick 2003). Micro-RNAs play important roles in Dicer- and Risc-mediated mRNA degradation or mRNA translation inhibition (Bartel 2004). This implies an ancient and effective RNAi immune function to combat viruses or viral-like agents such as transposons and retroposons (Moazed 2009; Jinek and Doudna 2009; Siomi and Siomi 2009; Obbard et al. 2009). Because micro-RNAs act on mRNAs, not on proteins, they are probably encoded by persistent nuclear DNA viruses (Cullen 2006, 2009). We will look at these competences later.

10.4.2 Former Competing Genetic Parasites Built Addiction Modules

The persistent status is the result of multiple colonization events into a host. This neutralizes former antagonistic and incompatible features of competing viral agents without harming the host (Ryan 2004, 2006, 2007). Most of the endogenous or exogenous inhabitants inherent to bacteria, protozoa, plants, animals and fungi are a complementary mix of formerly antagonistic viral features. They can still be identified today as toxin/antitoxin, restriction/modification-, insertion/deletion modules, i.e. complementary counterpart regulatory functions which do not harm the host (Villarreal 2005; Gerdes 2000; Makarova et al. 2006). As symbiotic neutralization and counterpart regulation, they represent new phenotypic features. One feature is regulated exactly by the antagonist according to developmental stages in the cell cycle, replication, tissue growth, or similar developmental contexts. Should this suppressor function become unbalanced, then the normally downregulated part may become lytic with potentially lethal consequences, as documented for symbiodinium and its major role in coral bleaching (Witzany 2007b).

10.4.2.1 The Symbiogenetic Lifestyle of Retroviruses

Retroviral activities in the persistent status are often characterized by features expressed only in the strict time window of a developmental process, such as axis formation, trophoplast formation, or the S phase of the cell cycle. In these highly specialized contexts they are replicated through signalling, which blocks the suppression of the replication process. After the function is fulfilled, a signal once again initiates suppressor function. Retroelements – with their (i) higher-order regulatory functions, (ii) capability for genetic creativity and (iii) capacity for innovation of new regulatory patterns and combinations – descended from retroviruses, which can easily be identified by their three essential parts *gag*, *pol* and *env* (Rashkova et al. 2002; Peterson-Burch and Voytas 2002; Tang et al. 1999a). Most endogenous retroviruses have been degraded into formerly connected domains, but they can still be recognized by retroposons or one of these three genes (Ryan 2006; Gao et al. 2003; Sfakianos and Hunter 2003; Ivanyi-Nagy and Darlix 2008), which means their formerly connected genomic content may be used by host organisms as single or networking modular tools for a variety of new regulatory functions. The *gag* gene encodes structural proteins, *pol* encodes enzymes such as reverse transcriptase, protease, ribonuclease and integrase functions, and *env* encodes envelope proteins, surface and transmembrane proteins and proteins causing host cell fusion and immunosuppression (Ryan 2007).

10.4.2.2 'Highways' that Play Important Roles in Persistence

We can find small DNA viruses as genetically stabilized and co-evolved persis-tent viruses that do not trigger and are not part of an immune response of the host organism. Their active role is regulated and depends on the cell cycle of the host in which they are transcribed, replicated and silenced again during a certain phase of the cell cycle. This means their persistent status is changed into an active role only during a strictly defined phase of the cell cycle in which they are needed for a specialized function in which they are still competent, most likely being adapted especially for this function. After fullfilling this function they disappear again. To ensure this function and its fine-tuned regulation they need a highly conserved regulative viral protein domain with characteristic host-interactions to use (manipulate) the host replication for their special needs. Their behavioral pattern is adapted to the host and circumvents the acute lytic phase (Shadan and Villarreal 1996).

We can find similar reproduction patterns in RNA viruses. Replication of retroviruses in eukaryotes depends on successful entry into the membrane of host cells (Nisole and Saib 2004). Some retroviruses circumvent this active entry of the cellular membrane in that they wait until the start of the cell cycle phase-dependent dissolution of the cell membrane during replication. Once the direct path to the nucleus is free, retroviral RNA is transported to the eukaryotic nucleus and integrates into the host genome. Through transcription of the host genome a complete viral RNA-genome is processed and transported out of the nucleus through the cytoplasm to the cell membrane. Here the reproduced viral genomes assemble and are encapsulated by viral gag encoded structural proteins and leave the host cell. Retroviral life integrates dozens of retroviral competences such as replication, transcription, translation, repair, trafficking to and from the nucleus, splicing, alternative splicing and 3'end processing. Transport to the nucleus and afterwards from the nucleus to the membrane again gives an overview about agent-driven evolution of eukaryotic cells if we think about the eukaryotic nucleus being an ancient dsDNA virus.

In contrast to these retroviral and often lytic replication cycles the overwhelming majority are non-lytic but persistent retroviruses. They infect the host organism and integrate their gag, pol, env functional parts into the host genome and adapt to the replication cycle of the host organism without leaving the host cells. In becoming part of the identity of the host genotype they change the genome formation and transfer a phenotype to the host that non-infected host genomes do not possess. We are still at the beginning of imagining how the persistent lifestyle of viruses plays a role in the evolution, development and genomic regulatory ratio of eukaryotes (Villarreal 2005).

This advantageous behavior of not leaving the host genome again but reaching a persistent status within the host genome (most probably by becoming part of an addiction module. i.e. a competing genetic settler that creates an equilibrium status balanced by the host immune system) can be better understood if we look at the patterns of retroviral movement during infection events.

10.4.2.3 The Kinesin/Dynein Addiction Module

Prokaryotic viruses tend to use pores during cell distruction and exit. Their entry pores are their tail plates, which can also be toxins. In eukaryotes, the most relevant pores are on the mitochondria, associated with apoptosis. However, eukaryotic DNA viruses (adenovirus) do often bind to nuclear pores and these are clearly also associated with microtubules that transport virus to the pore. This pore-building ability plays an evolutionary role in all tubular structures that connect multicellular tissues of eukaryotic organisms.

The alternative way of most eukaryotic viruses including retroviruses is to pass through the cell membrane of the host is by endocytosis followed by receptor binding (Greber 2002). A well-known behavioral motif then occurs: the release of its RNA or DNA into the host cell and its *immediate spread into smaller parts* such as reverse transcriptases and pre-integration-complexes. This is an important behavioral motif in relation to non-coding RNA transcript processing into small non-coding RNA species, as discussed later.

These smaller parts move on the 'highway' of the actin-based cytoskeleton and its microtubules in the direction of the nucleus (Greber 2005) by using the kinesin–motor protein superfamily (Vale 2003). After reaching the nuclear membrane these viral parts pass through the membrane through nuclear pores by using importine proteins (which consist of two subunits with complementary functions: an indicator for an ancient addiction module itself) which bind to a special recognition sequence.

Afterwards integrase (which is also produced and used by DNA viruses for the same purpose) integrates these viral parts into the genome. The integration process is not random but involves strict coherence to the syntax of the nucleotide sequences of the host.

If the host cell replicates these RNA viral parts are transcribed into DNA. As DNA sequences they pass through the membrane of the nucleus again to move towards the cell membrane. Again they use the 'highway' of microtubles but unlike earlier, they use dynein as the motor protein (Vale 2003). Interestingly the change in direction of the kinesin/dynein transporter proteins depends on suppression of antagonists of the dynein or the kinesin. This could be an indicator that both are part of an addiction module of former competing genetic parasites. Prior to becoming an internal part of cellular transport it could have been an external system for movement in and out of cells. And indeed there is a connection between these two motor proteins and retroviral (gag)-parts (Tang et al. 1999a; Cochrane et al. 2006).

When they reach the cell membrane the viral parts build a patch, via protein assembly on internal membrane, that is divided from the cell by exocytosis and produces its own capsule wherin the RNA genome matures. In other contexts similar proteins are involved in cytokinesis (Carlton and Martin-Serrano 2007) and most interestingly in neural and immunological synaptic communication (Dustin and Colman 2002; Piguet and Sattenaut 2004). Other retroviruses build their capsule not at the plasma membrane but at the centrioles (Sfakianos and Hunter 2003). Both sites of capsid building and transport depend on intact env and gag codings and the recycling of membrane parts (Cochrane et al. 2006). The gag-parts bind the kinesin motor proteins that are needed for mictrotubulin transport. The exact transport of the retroviral RNA through the cytoplasm depends on interactions with numerous host proteins that build the so-called RNA-Transport-Granulat (RTG) (Cochrane et al. 2006). This RTG is also present in nearly all cells such as fibroblasts, T-cells and epithelial cells (Shav-Tal and Singer 2005; Mingle et al. 2005).

Retroviral RNA editing for processing of extracellular viral parts is a very complex process: retroviral RNA editing functions in a similar way to cellular mRNA processing but is much more regulated by cis- and transactive mechanisms, which seem to be a former retroviral competence. Export of retroviral RNA out of the nucleus requires several RNA helicases such as RNA helicase A, DEAD box proteins DDX1 and DDX2 (Yedavalli et al. 2004; Reddy et al. 2000; Li et al. 1999; Tang et al. 1999b).

After retroviral RNA processing in the nucleus by (i) alternative splicing, (ii) 3'end processing and (iii) RNA transport from the nucleus through the cytoplasm on the microtubule highways, they reach the membrane where they assemble. Transport depends on intact microtubules (Basyuk et al. 2003). One part of the retroviral RNA is not spliced but is translated into structural proteins to form the capsules (Butsch and Boris-Lawrie 2002; Kaye and Lever 1999; Poon et al. 2002; Butsch and Boris-Lawrie 2000). The addiction module of the antagonistic motor proteins kinesin/dynein (see note above), which drive retroviral RNA transport, plays an important role in the cell division of eukaryotic cells (Cottingham and Hoyt 1997; Terada and Hirokawa 2000). Different kinesin proteins regulate (a) the movement (kip3p) but also (b) the direction/orientation of this movement (Kip3p) (Miller et al. 1998).

Interestingly, mitotic spindle processing without kinesin/dynein transcription does not function as well as positioning of the two poles and the segregation of the chromosomes in the ana-phase (Hoyt et al. 1992; Saunders et al. 1995; Hoyt et al. 1997; Haraguchi et al. 2006). Both motor protein families have genetic similarities with Archaea and Eubacteria, indicating their important roles in prokaryotes as well (Defeu-Soufo and Graumann 2004; Graumann 2004; Graumann and Defeu-Soufo 2004). Motor proteins and cytoskeleton interactions are very specific and also interconnect with the Golgi apparatus (Preuss et al. 2004; Lu et al. 2005).

10.5 Competent Regulators of Gene Expression

Non-coding RNAs that function in gene regulation coordinate and organize various actions such as chromatin modification and epigenetic memory, transcriptional regulation, control of alternative splicing, RNA modification and RNA editing, control of mRNA turnover, control of translation, and signal transduction (Bird et al. 2006; Amaral et al. 2008).

In contrast to former opinions about the expression levels of genomes it is now increasingly clear that most eukaryotic genomes are highly expressed and a great abundance of non-coding RNAs with regulatory functions is transcribed. Most of these non-coding RNAs are alternatively spliced and divided into smaller RNAs that are integral parts of ribonucleo-protein complexes. They regulate nearly all aspects of gene regulation. Small RNA species include microRNAs, small interfering RNAs, small nuclear RNAs, small nucleolar RNAs and transfer RNAs (Yazgan and Krebs 2007).

Although recent research has tried to evaluate the enormous regulatory networks of small RNAs, the role of thousands of longer transcripts is not yet clear. We know that they play important roles in histone modification, methylation, i.e. epigenetic control of developmental processes such as the mammalian HOX clusters (Amaral et al. 2008), and also transcriptional interference, promoter inactivation and effects on enzymatic pathways. Interestingly these large non-coding RNAs are found as interlacing and overlapping sense and antisense transcripts derived from introns or intergenic regions, which means that they stem from the preferred life habitat of persistent viral settlers. That they may be of viral descent is indicated by their developmental stage- and tissue-specific expression patterns, which are typical habits of persistent viral settlers. Similar to their smaller relatives they are involved in the formation of ribonucleoprotein complexes.

10.5.1 Identification and Regulation by microRNAs and siRNAs

Small non-coding RNAs also share a special competence for epigenetic regulation of gene expression (Chu and Rana 2007) and are derived from repetitive genomic sequences (Farazi et al. 2008). Repetitive genomic sequences indicate descent from retroviral infection events (Witzany 2006). The capacity for epigenetic regulation of gene expression includes the 'recognition' (identification) of specific sequences in other nucleic acids and is common to RNAs (Filipowicz 2000), especially small nuclear RNAs and tRNAs which (i) identify splice junctions in both pre-mRNAs and codons and (ii) process both the subunits of the spliceosome and the ribosome (Mattick 2003). This implicates their capacity for self/non-self distinction as well as for identifying the molecular syntax. If one of these does not function, i.e. error or damage occurs, the regulation or structural features of small non-coding RNAs do not function.

Maybe this is a key feature of non-coding RNAs i.e. they share an analog/digital language-competence (St.Laurent and Wahlestedt 2007), such that in their secondary molecular structure they can act as molecular adaptors to the protein world, whereas their nucleotide word order seems to be digitally structured information. Their ability to edit the molecular syntax of genetic texts according to different needs is exemplified by recent research on the functions of learning and memory in mammalian neuronal networks (Presutti et al. 2006; Mehler and Mattick 2007). This is possible in that RNA editing alters transcripts from loci encoding proteins involved in neural cell identity, resulting in DNA recoding (Mattick and Makunin 2006; Mattick and Mehler 2008).

The action of endogenous RNA species such as microRNAs, small interfering RNAs and piwi RNAs in animals and plants is a mediating process in that they guide the binding of protein complexes to specific nucleic acid sequences (Ambros and Chen 2007). There action starts both as a regulatory process at the transcriptional level, e.g. by endogenous siRNAs (Watanabe et al. 2008) when the action potential is activated out of the 'evolutionary protocols' fixed in the DNA storage medium, or at the post-transcriptional level in that they stabilize messenger RNA and translation into proteins (Chen and Rajewsky 2007). This means that small non-coding RNAs do not solely mediate the transfer of genetic information from DNA to protein but also act as sequence-specific regulators in the expression of other RNA transcripts and, interestingly, in silencing specific transposons (He and Hannon 2004; Bartel 2004; Chu and Rana 2007).

Control patterns include mRNA degradation (by siRNAs), translational repression (by miRNAs), heterochromatin formation and transposon control (by piwiRNAs) (Chu and Rana 2007). Endogenous miRNAs and siRNAs share biogenesis and can perform interchangeable functions (Doench et al. 2003).

They cannot be distinguished by their chemical composition or their action. But they differ in their production pathways: (i) miRNAs derive from genetic sequences that are different to known genes, siRNAS derive from mRNAs, transposons, viruses, (ii) miRNAs are processed out of transcripts that can form RNA hairpins, siRNAs are processed from long bimolecular RNA duplexes, (iii) miRNAs are always conserved in related organisms, endogenous siRNAs are rarely conserved, (iv) miRNAs are produced from genes that are specialists in silencing of different genes, siRNAs are typically auto-silencing, such as viruses, transposons and repeats of centromeres (Bartel 2004). SiRNAs are expressed from extended double-stranded regions of long inverted repeats and can inhibit the expression of nearly any target gene in response to double-stranded RNA and have a very efficient and ancient immune function against genetic parasites (Fire 2005; Sontheimer and Carthew 2005). This functions by identifying foreign RNA sequences and inhibiting their replication (Elbashir et al. 2001). The ancient RNAi immune function is based on self/non-self identification competence (Fire 2005). Many of these elements are retroposons or transposons and are encoded in the repetitive sequences of the genome (Sontheimer and Carthew 2005; Jurka et al. 2007) and therefore are clearly of viral origin.

MicroRNAs are single-stranded RNAs and are generated from endogenous hairpin transcripts of 70 nucleotide precursor miRNAs (Kim 2005). The transcription of this pre-miRNA is processed by RNA polymerases pol II and pol III. Whereas pol II produces the messenger RNA, small nucleolar and small nuclear RNAs of the spliceosome pol III produce shorter non-coding RNAs such as tRNAs, some rRNAs and a nuclear RNA that is part of the spliceosome (Bartel 2004). They control not only developmental timing, hematopoiesis, organogenesis, apoptosis, and cell proliferation but also fat metabolism in flies, neuronal patterning in nematodes and control of leaf and flower development in plants (Bartel 2004). Most of them are processed out of introns!

It is predicted that every metazoan cell type at each developmental stage has a distinct miRNA expression profile (Bartel 2004). The most characteristic differences of miRNAs acting in plants and animals are found in the stem loop.

MicroRNAs (19–25 nucleotides in length) as well as small interfering RNAs (21–28 nucleotides in length) seem to have descended from transposable elements with an inherent regulatory ratio on gene regulation that is fulfilled by a variety of small interfering RNAs or microRNAs which act in a coordinated manner in that they share a division of labor in hierarchical steps of suppression and amplification. This is indicated in transposable elements that encode both siRNAs and miRNAs (Piriyapongsa and Jordan 2008). They can be found in intronic regions and build stem loop structures (hairpins) as a common feature of active RNA species like ribozymes. The defense mechanism of host genomes against transposable element-invaders through siRNA evolved into miRNAs with a new regulatory complexity and a new phenotype. First evolving as an immune function, it was later co-opted as a tool for complex regulatory pathways for host gene expression (Piriyapongsa and Jordan 2008). This co-option of identical competences for different purposes seems to be a common evolutionary pattern for regulatory controls that can be flexibly altered and re-arranged to cause phenotypic variation without altering basic components (Mattick and Gagen 2001).

Micro RNAs are acknowledged as key regulators of gene expression. This means that dysfunctions of these regulatory mechanisms may lead to dysregulated genes with a cascade of disease-causing consequences (Calin and Croce 2006). This means that the smallest RNAs are the basis of gene regulation, although in their original function they had an immune function against viruses and similar agents. Later on their function was adapted for eukaryotic gene expression. Sometimes these new functions overlap with old features when virus infection is acute (Quellet et al. 2006) and the balanced (persistent) status is disturbed. The eukaryotic signalling pathways that act during gene expression or DNA replication and the role of transport from the cytoplasm through nuclear pores into the nucleoplasm, genetic expression and retransport into the cytoplasm and the role of membranes, show that viruses had and still have manipulatory abilities in host genomes and host cells (Miller and Krijnse-Locker 2008). In particular, the ability of persistent invaders to self-splice out before export to the cytoplasm – an ability derived from the division of transcription and translation through nuclear membranes with pores – produces an analog protein-coding dataset ready for translation and determines the identity of the host without harming its reproductive cycle.

10.5.2 Non-coding RNAs act as Ribonucleoproteins

All these RNA interactions and the great variety of functions as preconditions of eukaryotic complexity include gene–gene interactions as well as the integration and regulation of most of the gene activities on different levels, such as chromatin structure, DNA-methylation, transcription, RNA splicing, RNA translation, RNA stability and RNA signalling. This means that most of the functions of gene control that we currently know (Mattick 2001, 2003, 2006, 2007) descended from retroelements. One important feature is that non-coding RNAs function as ribonucleoproteins and not as naked RNAs (Matera et al. 2007).

Some of them are small nuclear and small nucleolar RNAs (see below). Interestingly the nucleolus plays an important role with similar performance to other endogenous retroelements during a small time window in the cell cycle. Nucleoli appear during the inter-phase of mammalian cells and are the locus of ribosome processing. Nucleoli also control regulation of the cell cycle (Hiscox 2002). In general, non-coding RNAs play important roles as intermediate products in the function and structure of the host genome, e.g. in the reproductive cycle of host cells from the start to the end of the S-phase. This is a common feature of persistent endogenous retroviruses that adapted to the host genome as an addiction module. They now represent a phenotypic function of the host organism as neutralized (balanced) former competing viral agents. If one of these agents is damaged or its regulation becomes unstable the counterpart may still become virulent with potential disease-causing consequences for the host organism (Witzany 2006).

Regulatory functions are restricted to specific phases of the cell cycle. One antagonist is suppressed and therefore the other can be produced at increasing rates according to the phase of the cell cycle, e.g. production of tubulin and/or nanotubulin structures (Rustom et al. 2004) or production of proteins with a variety of functions in cell division. At the end of this specific phase of the cell cycle the suppression of the antagonist ceases, mediated by specific signalling processes, and the antagonist then suppresses the increased production of its counterpart. This is the precondition for terminating its function in this phase of the cell cycle. These interactions represent a kind of universal module with key regulatory function. Non-coding RNAs are involved in nearly all of these key functions of cellular activities in all domains of life (Prokaryotes, Eukarya, Archae).

10.5.3 Small Nuclear and Small Nucleolar RNAs

Most of these functions are fulfilled by non-coding RNAs, which act as binding partners to ensure the correct position of the nucleic acid target molecule for its enzymatic functions. Normally this process works as Crick-Watson base-pairing which includes a lot of proteins. These interconnected networks between non-coding RNAs and proteins are termed the ribonucleoproteins (RNPs) (Matera et al. 2007).

Some of these non-coding RNPs are small nuclear (sn) and small nucleolar (sno) RNPs. They are competent in multiple functions, many of them for intracellular transport and motility. In human cells they are encoded within introns. Both act in a *complementary* fashion: snRNPs retrieve the snoRNAs from the introns. On the other hand snoRNAs are required by the snRNPs for post-transcriptional modifications. Their interdependency seems to be typical for viral derived addiction modules. Both snRNPs and snoRNPs depend on stable but inactive pre-RNPs, which only can mature if they are located far from their later active position (Matera et al. 2007).

In most vertebrates the snoRNAs come from introns of pre-RNA transcripts. These snoRNAs are integrated into a complex structure by endonucleases, exonucleases and helicases (Zemann et al. 2006; Richard et al. 2006). Interestingly in the yeast *Saccharomyces cerevisiae* most snoRNAs are not encoded in introns (Filipowicz 2000). This could be an indicator of an earlier evolutionary phase in which snoRNAs had not yet become endogenous by an infection event by retro-agents.

Most of the snoRNAs of vertebrates that are encoded in their introns are transcribed by polymerase II and produced after splicing through exonucleolytic trimming. They are highly lineage-specific and form a separate family of mobile genetic elements (Weber 2006).

Genes coding for snoRNAs switch on host genes by retroposition. They also play important roles in excision and integration of specific sequences. They are as important as Alu, SINEs and LINEs (Weber 2006). Interestingly, the best-known mobile genetic elements are integrated into snoRetroposons. The insertion is characterized by their individually different target site duplication-repeats e.g. full-length LINEs, i.e. ALUs in primates. snoReverse Transcriptases (snoRTs) are also inserted into other mobile elements as well as into DNA-transposons (Weber 2006). Retroelements seem to be *compatible with each other* and are integrated into small nucleolar reverse transcriptase sequences, each of them with its individual target-site duplication-repeat (Weber 2006). This could be an indicator of identification processes and identity-modules. Many of the snoRNA genes are retroposons with retroviral ancestry (Weber 2006). snoRTs that are part of the introns of host genes can be used as RNAs with novel functions. After a retroposition event they can silence the snoRNA copy of their parents and lead to a new function of the new snoRNA (Weber 2006). This seems to be a common feature of endogenous retroposons and transposons in that these genetic invaders are driving forces of genomic creativity: the same competences in different contexts may lead to new functions and phenotypes.

Transport of small nucleolar RNPs is characerized by both nuclear and cytoplasmic cycles. Cajal bodies play important roles in intranuclear transport as centers for RNP- assembly, -transport, -modification and editing. Their production is a complex highly coordinated and conserved process. When the small nuclear RNPs arrive in the nucleoplasm they spread over the whole interchromatin space. Newly generated RNPs are assembled in the Cajal bodies before they are integrated into the nucleolar subdomains fibrallin and interchromatin clusters. This seems to be an indicator that they are processed and modified within the Cajal bodies and specialized (informed) for specific functions (Matera et al. 2007). Beside the function as *de novo* assembling-factors of RNPs, multicompetent Cajal bodies play an important role in modification and recycling processes of U4/U6 snRNP-complexes that are remnants of splicing processes (Matera et al. 2007). Cajal bodies are nuclear organelles with high motility and important roles in splicing, ribosome-processing and transcription (Platini et al. 2002). It has been observed that Cajal bodies move from one end of the nucleus to the other, assemble with other Cajal bodies or divide themselves into smaller ones (Tomlinson et al. 2006). These endogenous retroelements are still important agents in natural genetic engineering and natural genome editing.

10.5.4 Currently Identified Roles of Small Nucleolar RNAs

Small nucleolar RNAs seem to derive from a very ancient group of endogenized RNA viruses with a wide variety of functions such as 2'-O-methylation, pseudouridylation of many classes of RNAs, rRNA processing and synthesis of telomeric DNA (Kiss 2002).

Today we know two families of snoRNAs: the C/D and the H/ACA RNAs. They are produced in the nucleolus and have a variety of functions within the nucleolus but also out of the nucleolus as a recreation center for special substances. Eukaryotic cells have several dozen species of snRNAs and 200 known snoRNAs (C/D and H/ACA RNAs). These RNAs are one of the most diverse transacting RNAs currently known. They are available not only in Eukaryotes but also in Archaea. These snoRNPs (C/D H/ACA) share important functions such as: protein translation, mRNA splicing, genome stability, ribosome function, and modifications in snRNAs of eukaryotes, tRNAs in Archaea and neuronal mRNAs in mammals. One kind of H/ACA RNA, telomerase RNA, is needed for telomere production (Kiss et al. 2002).

Most of the known snoRNPs guide modifications of other ncRNAs. Both motifs are simple and very ancient and are part of the telomerase RNA (Clouet d'Orval et al. 2001; Wang and Meier 2004). Depending on their function, C/D and H/ACA finally orientate to nucleoli, Cajal bodies or telomeres (Matera et al. 2007).

We know that one H/ACA RNP functions in vertebrate telomere synthesis. Telomerase RNA and its protein partner, telomerase reverse transcriptase, are both target orientated and strictly regulated. Telomerase enzymes are active in telomere elongation only in the S-phase of the cell cycle. Interestingly telomerase does not elongate telomeres continuously but at different times with interruptions in between cellular subcycles, especially in the shorter ones. The 3' region of the endogenous human telomerase RNA possesses all the structural features of H/ACA boxes of snoRNAs. Only in the S-phase of the cell cycle do both telomerase RNA and telomerase reverse transcriptase move towards telomeres, otherwise they are in different (waiting?) positions (Witzany 2008a). The accumulation of human telomerase RNA in nucleoplasmic Cajal bodies in certain cancer cells occurs only during the S-phase of the cell cycle when telomerases are generated (Jady et al. 2004). This means that the accumulation process is cell cycle dependent and time limited. Telomerase transport in the S-phase of the cell cycle is advantageous because telomerase activity in eukaryotes is limited to chromosome replication and suppresses destructive functions of telomerases at non-telomere elongation locations such as chromsome repair and repair of double-strand breaks (Tomlinson et al. 2006).

Cajal bodies in particular play important roles in the generation and function of telomerase RNPs. Human telomerase RNA has a H/ACA motif that it shares with small Cajal body RNAs and a wide variety of snoRNAs. The snoRNA family regulates modifications and cleavage of ribosomal RNAs in the nucleolus, and the small Cajal body RNA family regulates modifications of small nuclear RNAs within the Cajal body (Fu and Collins 2006).

Besides the nucleolus the Cajal bodies are the best investigated nucleoplasmatic organelles. They are enriched by spliceosomal RNPs and nucleolar RNPs. Especially in the interphase many nucleolar intermediates emerge and play important roles in transport and modification of a variety of cellular RNAs (Darzacq et al. 2002). C/D and H/ACA boxes guide the post-transcriptional pol-II specific spliceosomal snRNAs via pseudouridylation, which is characteristic of post-transcriptional RNA modifications in eukarya and archaea and generally plays an important role in the correct functioning of cellular RNAs. These guide RNAs share a Cajal body-specific localization signal, the CAB Box (Richard et al. 2003). It is suggested that pseudouridylation guide-RNPs play important roles in the processing of rRNA and in function-control/regulation of telomerases in eukaryotes. Ribosomal RNAs play important roles and therefore are DNA-encoded in hundreds of transcription units. These units are organized in large tandem arrays. In active synthesis these rRNA loci form nucleoli with important roles in evolution and development. These rRNA loci are habitats for mobile elements like retroposons (Eickbush und Eickbusch 2007). Pseudouridylation pockets also seem to direct snoRNAs to the nucleolus, which means that they have identity functions (Narayanan et al. 1999).

In yeast, telomerase is not associated with H/ACA snoRNPs but with SM snRNPs, which demonstrates that telomerase is generated and regulated in different organisms in different ways (Matera et al. 2007).

10.5.5 The tRNA Consortium

Additionally to the ribosomal rRNAs, tRNA is one of the most important RNAs and is essential for the replication of both RNA genomes and DNA genomes. It seems to be a very ancient competence because it functions in a wide variety of contexts in quite different ways. Investigations of the variety of functions of tRNA show that it seems to be an addiction module-like association of different predecessors dating to the RNA world. The 3' end of the tRNA structure could be the start of an ancient replicase. Replicase is the CCA-adding enzyme that is necessary to complete nucleotides that were lost because of incorrect starts. This CCA-adding activity has been the first telomerase function with its function of completing lost nucleic acid end-sequences during replication. If it is really as old as suggested it must be present in all three domains of life: archaea, bacteria, eukarya. These CCA-adding enzymes are found in all three domains of life, are part of the same nucleotidyltransferase superfamily and are very similar in their function (Maizels and Weiner 1999; Maizels et al. 1999).

The genomic tag hypothesis is another part of the puzzle over the role of non-coding RNA abilities. It is suggested that the one half of tRNA evolved to mark single-stranded RNA genomes for replication in the early RNA world. The second half of the tRNA evolved separately as a primer of templated protein synthesis and started the RNP-world. Both parts derived from independent RNA agents, which together built a kind of addiction module. Then this module became involved in translation from RNA template into protein. That tRNA plays important roles in the replication of single-stranded RNA viruses of bacteria, plants and mammals, replication of duplex DNA plasmids of fungal mitochondria, retroviral replication, and also replication of present chromosomal telomeres, is less noticed nowadays (Maizels and Weiner 1999; Maizels et al. 1999).

Last but not least let's have a look at one of the most prominent and ancient natural genome editing competences: reverse transcriptases such as telomerases that function in telomere-maintenance are non-coding retroagents. This indicates that their ancestry is of retroviral origin (Witzany 2008b). The vast majority of retroelements use tRNAs or RNAs with strong secondary structures to process reverse transcription. Interestingly there is a similarity between the function of tRNA in the production of proteins from RNA information, and reverse transcriptase, an enzyme that is crucial for turning RNA into DNA (Presutti et al. 2006).

Retroposition – a billions of years-old process – still plays important roles in building the structure and function of genomes in a continuing interaction between host-genomes and the colonizing life strategies of mobile genetic invaders – an everlasting evolution-driving process of rearrangement, renovation and innovation

(Brosius 2005). The ancient and prominent role of reverse transcriptases I have outlined in Chapter 9.

10.6 Reciprocal Interacting Agents

In contrast to DNA with its stable features and enormous information storage potential, RNA is involved in the active parts of copying and coding processes, as demonstrated in new sequence generation, replicative processes, gene invention and higher order regulations in all key processes of life. Prior to the evolution of cellular life it is proposed that very simple structured RNA (pre-RNAs) started by growing through base-pairing mechanisms without coding features. The selection of an RNA population a direct product of error prone *unedited RNA replication* is known as the quasispecies theory.

Growth by base-pairing mechanisms is different to the growth of primitive secondary structures of single-stranded RNAs, which can stabilize and replicate themselves as hairpin/stem loop structures with inherent coding capabilities. When coding began, catalytic functions connected with syntactic rule-ordered information enabled these simple molecules to act as *semiotic agents*: they became capable of generating nucleic acid sequences with a functional meaning which had to be recognized, identified and interpreted correctly in the situational context, with a combinatorial pattern of the base pairs that differed from the diverse features inherent in non-self agents of the same structure, i.e. self/non-self identification (Villarreal 2009). Biotic competences differ from abiotic interactions, because in contrast to the latter, biotic competences which became active may even fail. These abilities are still evident in the t-loop structure of tRNAs, a variety of ribozymes and self-enforcing RNAi loops, which couple heterochromatin assembly to siRNA production (Sugiyama et al. 2005).

The high density of early RNA life led to competing situations in which it was an advantage to escape into DNA informational storage and protein-based cellular life as outlined in Chapter 8 (Witzany 2008b). Protein- and DNA-invention were a prerequisite for the evolution of divergence and the variety of life because all evolutionary inventions could be stored as evolutionary protocols (Vetsigian et al. 2006) in this stable storage medium. The information content of the human genome is comparable to an archive of 5000 books with 300 pages each.

It appears that all the detailed steps of evolution stored in DNA that are read, transcribed and translated in every developmental and growth process of each individual cell depend on RNA-mediated processes, in most cases interconnected with other RNAs and their associated protein complexes and functions in a strict hierarchy of temporal and spatial steps. It is clear that this regulatory order could not evolve by chance or that it represents solely a randomly-derived mixture of nucleotides, but that it is composed of individual functions and integration into one developmental target, in strict coherence to the syntax of the nucleic acid language.

Today we are beginning to realize the degree of abundance and variety of RNA species with their different, sometimes complementary and competing roles in all key processes of life. RNAs play complementary roles in information processing and regulation (Kenzelmann et al. 2006). In most cases they have an inheritable status, being integrated in the genome of organisms, and are termed endogenous. In other cases they are ancient individuals living in the cytoplasm of cells as persistent non-lytic parasites similar to DNA-settlers, with important endosymbiotic roles. Their relation to viruses is close and some virologists consider an evolutionary tree of RNA-species and RNA viruses. Interestingly, some DNA viruses have other features (linear chromosome, telomere ends, intron-like structures) that indicate a different origin with ancient roots comparable to RNA viruses. These features connect the evolutionary roots of Archaea and Eukaryotes, because ancient dsDNA viruses have similar features in viruses of Archaea and the eukaryotic nucleus. A special feature is the RNA proofreading and repair ability of RNA polymerases, which would be the precondition of an RNA genome in the early RNA world because of the relatively unstable RNA structures (Jeffares et al. 1998; Poole et al. 1998; Poole and Logan 2005). On the other hand, this instability is a necessary precondition for the high productivity of different RNA sequences with the rapidly adapting features necessary in the early RNA world, i.e. a large variety of different RNA identities.

In particular, the mode of replication of eukaryotes shows a wide variety of hierachical ordered processes which each depend on signalling processes to indicate the successful termination of the preceding process. A temporal order exists, with time windows processed by different process-design connected with the cell cycle, along with transport systems of signals and complex messages, agents, co-agents and helpers such as ancillary proteins and a network of interwoven regulatory elements which suppress or amplify the start and stops of semioses, production, regulation and the whole toolbox of natural genetic engineering. We now know that all these processes involve RNAs, which become active via transcription out of the DNA storage medium prior to translation into proteins.

During translation from digital DNA storage into the analog code of protein-language it is interesting that DNA information contains multiple RNA- and protein-meanings, i.e. from the same genetic data set it is possible to transcribe multiple RNA species or translate a variety of proteins according to the higher order regulations inherent in epigenetic control and/or transcriptional, pre- or post-transcriptional modification targets. Additionally in eukaryotes we find some non-coding RNA species such as small nuclear, small nucleolar, small Cajal body, and small interfering RNAs. There are some indications that non-coding DNA plays a role in the *de novo* generation of genes (Levine et al. 2006).

Nearly all genetic and genome editing processes involve RNAs, which in most cases function as a network. Most of the functions are performed, conducted and regulated by non-coding RNAs, which are encoded in intronic DNA in most cases with repetitive syntax. Most of their functions are active only in an intermediate stage of RNA processing, which is regulated by strict starts and stops that are signal mediated. Error in these regulations is interconnected with organismic diseases.

Non-coding RNAs are similar to all kinds of mobile genetic elements such as LTRs, non-LTRs, SINEs, LINEs, snoRNAs, and snRNAs, all of persistent viral origin. They are integrated via addiction modules, i.e. an agent/antagonist relationship between competing genetic settlers that are neutralized and balanced by their antagonism and the immune system of the host. This is an advantage for both the host, which attains a new genetic phenotype that non-infected relatives do not possess, and the viral settlers, which both survive and co-evolve within a new genomic habitat.

Transfer RNA (tRNA) and ribosomal RNA (rRNAs), with their key functions in protein translation, represent such addiction modules, containing several subunits, each of them necessary for function. The whole cell cyle of eukaryotes is regulated by these non-coding RNAs.

Obviously DNA functions as both a relatively stable information storage medium – an evolutionary protocol to fix advantageous innovations – and as a comfortable habitat for persistent genetic settlers. If all the RNA capabilities derive from viruses or similar agents which compete in the available global pool of organismal genomes – viruses and their relatives are ten times more abundant than cellular genomes – then only those that give their hosts an advantageous genomic identity that is able to ward off an abundance of competing genetic settlers will survive. They must be able to build addiction modules (genetic and genomic innovations) together with the host immune system, each of them a unique culture-dependent habitat. Only then will the survival of both the genetic settlers and their host populations be likely.

10.7 Conclusion

At the genomic level we can identify a remarkable process of change from a mechanistic view to the perspective of non-mechanistic genetic content processing. If we look at the current knowledge of hierarchical and temporal order of single steps and substeps in replication and transcription processes there must be natural genome editing agents that are competent both in generation of meaningful nucleotide sequences and in the use of these sequences according to different needs such as integration, modification, recombination, and extraction into pre-existing genetic texts. As we will see, in this respect DNA is not only an information storing archive but a life-habitat for linguistically competent RNA-agents of viral or subviral descent. These agents are competent in almost error-free editing of nucleotide sequences according to combinatorial, context-sensitive and content-specific rules. They even generate nucleotide sequences *de novo*. They are also able to generate new rules of use for nucleic acid sequence modules by rearrangements in the higher order regulatory network of non-coding domains. Thus ancient sequences of the DNA storage medium may be used as modular tools in a wide variety of different contexts for new functions, made possible through the different meaning of syntactically identical sequences.

Competent agents in nucleic acid-language are not solus ipse agents but are competent as mutual or parasitic 'swarms' or 'clouds', most of them RNA-based communities that share these competences. Their competence is a communal one, each of them being capable of self and non-self identification. The interactive competence of a community enables each individual to be competent. If we look at interacting communities such as ribosomes and spliceosomes (each containing subunits without which they cannot function) we see their shared competence. If we look at the hierarchical processes of gene expression, transcription, RNA processing, mRNA and tRNA transport for translation we can also see reciprocal acting agents. From the virus-first perspective they are now mutually interacting but may derive from formerly competing agents. mRNA and tRNA maturation in eukaryotes in particular also seem to reflect communal processing.

Formerly competing agents have reached an equilibrium status balanced by the immune response of the infected host to achieve a persistent life-style in the host genome e.g. toxin/antitoxin-, insertion/deletion- or even restriction/modification modules. The number of reciprocal interacting agents represented by e.g. ribosomes, spliceosomes or even the consortium which cooperates as the adaptive immune system (Villarreal 2009) ranges from a few to hundreds and thousands. In the latter case communal agents interact in DNA rearrangements with enormous consequences for many protein-based products that play important roles in immune functions.

This view could change the construction of research projects, i.e. shifting the focus from mutational (random) changes of nucleotide sequences to investigating nucleotide sequences from the perspective of viral derived sequences that now play important roles in the regulation of cellular functions. Their status within one of many addiction modules can be changed by non-beneficial circumstances for the cell (e.g. stress) and they may become lytic again, resulting in a wide variety of diseases.

References

Amaral PP, Dinger ME, Mercer TR et al. (2008) The eukaryotic genome as an RNA machine. Science 319:1787–1789

Ambros V, Chen X (2007) The regulation of genes and genomes by small RNAs. Development 134:1635–1641

Bapteste E, Charlebois RL, MacLeod D et al. (2005) The two tempos of nuclear pore complex evolution: highly adapting proteins in an ancient frozen structure. Genome Biol 6:R85

Bartel DP (2004) MicroRNAs: genomics, biogenesis, mechanism and function. Cell 116:281–297

Basyuk E, Galli T, Mougel M et al. (2003) Retroviral genomic RNAs are transported to the plasma membrane by endosomal vesicles. Dev Cell 5:161–174

Bell PJL (2001) Viral eukaryogenesis: was the ancestor of the nucleus a complex DNAVirus? J Mol Evol 53:251–256

Bell PJL (2006). Sex and the eukaryotioc cell cycle is consistent with a viral ancestry for the eukaryotic nucleus. J Theor Biol 243:54–63

Bird CP, Stranger BE, Dermitzakis ET (2006) Functional variation and evolution of non-coding DNA. Curr Opin Genet Dev 16:559–564

Brosius J (2005) Echoes from the past - are we still in an RNP world? Cytogenet Genome Res 110:1–4
Butsch M, Boris-Lawrie K (2000) Translation is not required to generate virion precursor RNA in human immunodeficiency virus type 1-infected T-cells. J Virol 74:11531–11537
Butsch M, Boris-Lawrie K (2002) Destiny of unspliced retroviral RNA: ribosome and/or virion? J Virol 76:3089–3094
Calin GA, Croce CM (2006) MicroRNA signatures in human cancers. Nat Rev Cancer 6:857–866
Carlton J.G. & J. Martin-Serrano (2007) Parallels between cytokinesis and retroviral budding: a role for the ESCRT machinery. Science 316:1908–1912
Chen K, Rajewsky N (2007) The evolution of gene regulation by transcription factors and microRNAs. Nat Rev Genet 8:93–103
Chu CY, Rana TM (2007) Small RNAs: regulators and guardians of the genome. J Cell Physiol 213:412–419
Claverie JM (2005) Fewer genes, more noncoding RNA. Science 309:1529–1530
Clouet d'Orval B, Bortolin ML, Gaspin C et al. (2001) Box C/D RNA guides for the ribose methylation of archaeal tRNAs. The tRNATrp intron guides the formation of two ribose-methylated nucleosides in the mature tRNATrp. Nucleic Acids Res 29:4518–4529
Cochrane AW, McNally MT, Mouland AJ (2006) The retrovirus RNA trafficking granule: from birth to maturity. Retrovirology 3:18
Cottingham FR, Hoyt MA (1997) Mitotic spindle positioning in saccharomyces cerevisiae is accomplished by antagonistically acting microtubule motor proteins. J Cell Biol 138: 1041–1053
Cullen BR (2006) Viruses and microRNAs. Nat Genet 38:25–30
Cullen BR (2009).Viral and cellular messenger RNA targets of viral microRNAs. Nature 457: 421–425
Darzacq X, Jady BE, Verheggen C et al. (2002) Cajal body-specific small nuclear RNAs: a novel class of 2'-O-methylation and pseudouridylation guide RNAs. EMBO J 21:2746–2756
Defeu-Soufo JH, Graumann PL (2004) Dynamic movement of actin-like proteins within bacterial cells. EMBO Rep 5:789–794
Doench JG, Petersen CP, Sharp PA (2003) siRNAs can function as miRNAs. Genes Dev 17: 438–442
Dustin ML, Colman DR (2002) Neural and immunological synaptic relations. Science 298:785
Eickbush TH, Eickbush DG (2007) Finely orchestrated movements: evolution of the ribosomal RNA genes. Genetics 175:477–485
Elbashir SM, Lendeckel W, Tuschl T (2001) RNA interference is mediated by 21- and 22-nucleotide RNAs. Genes Dev 15:188–200
Farazi TA, Juranek SA, Tuschl T (2008) The growing catalogue of small RNAs. Exiqon. The 2008 Collection Booklet
Filipowicz W (2000) Imprinted expression of small nucleolar RNAs in brain: time for RNomics. Proc Natl Acad Sci U S A 97:14035–14037
Fire A (2005) Nucleic acid structure and intracellular immunity: some recent ideas from the world of RNAi. Q Rev Biophys 38:303–309
Forterre P (2001) Genomics and early cellular evolution. The origin of the DNA world. Comptes rendus de l'Académie des sciences. Série 3, Sciences de la vie 324:1067–1076
Forterre P (2002) The origin of DNA genomes and DNA replication proteins. Curr Opin Microbiol 5:525–532
Forterre P (2005) The two ages of the RNA world, and the transition to the DNA world: a story of viruses and cells. Biochimie 87:793–803
Forterre P (2006a) The origin of viruses and their possible roles in major evolutionary transitions. Virus Res 117:5–16
Forterre P (2006b) Three RNA cells for ribosomal lineages and three DNA viruses to replicate their genomes: a hypothesis for the origin of cellular domain. Proc Natl Acad Sci U S A 103: 3669–3674

Fu D, Collins K (2006) Genes Dev 20:531–536
Gao X, Havecker ER, Baranov PV et al. (2003) Translational recoding signals between gag and pol in diverse LTR retrotransposons. RNA 9:1422–1430
Gerdes K (2000) Toxin-antitoxin modules may regulate synthesis of macromolecules during nutritional stress. J Bacteriol 182:561–572
Gorinsek B, Gubensek F, Kordis D (2004) Evolutionary genomics of chromovirus in eukaryotes. Mol Biol Evol 21:781–798
Graumann PL (2004) Cytoskeletal elements in bacteria. Curr Opin Microbiol 7:565–571
Graumann PL, Defeu-Soufo HJ (2004) An intracellular actin motor in bacteria? Bioessays 26:1209–1216
Greber UF (2002) Signalling in viral entry. Cell Mol Life Sci 59:608–626
Greber UF (2005) Viral trafficking violations in axons: the herpesvirus case. Proc Natl Acad Sci 102:5639–5640
Haraguchi K, Hayashi T, Jimbo T et al. (2006) Role of the kinesin 2 family protein, KIF3, during mitosis. J Biol Chem 281:4094–4099
He L, Hannon GJ (2004) MicroRNAs: small RNAs with a big role in gene regulation. Nat Rev Genet 5:522–531
Hiscox JA (2002) The nucleolus – a gateway to viral infection. Arch Virol 147:1077–1089
Hoyt MA, He L, Loo KK, Saunders WS (1992) Two Saccharomyces cerevisieae kinesin-related gene products required for mitotic spindle assembly. J Cell Biol 118:109–120
Hoyt MA, Hyman AA, Bähler M (1997) Motor proteins of the eukaryotic cytoskeleton. Proc Natl Acad Sci U S A 94:12747–12748
Ivanyi-Nagy R, Darlix JL (2008) Ancient retrotransposons as possible remnants of the primitive RNP world. In: Gage FH, Christen Y (eds) Retrotransposition, Diversity and the Brain. Springer, Berlin/Heidelberg
Jady BE, Bertrand E, Kiss T (2004) Human telomerase RNA and box H/ACA scaRNAs share a common Cajal body-specific localization signal. J Cell Biol 164:647–652
Jeffares DC, Poole AM, Penny D (1998) Relics from the RNA world. J Mol Evol 46:18–36
Jinek M, Doudna A (2009) A three-dimensional view of the molecular machinery of RNA interference. Nature 457:405–412
Jurka J, Kapitonov VV, Kohany O et al. (2007) Repetitive sequences in complex genomes: structure and evolution. Annu Rev Genom Human Genet 8:241–259
Kaye JF, Lever AM (1999) Human immunodeficiency virus types 1 and 2 differe in the predominant mechanism used for selection of genomic RNA for encapsidation. J Virol 73:3023–3031
Kenzelmann M, Rippe K, Mattick JS (2006) RNA: networks & imaging. Mol Syst Biol 2:44
Kim VN (2005) MicroRNA biogenesis: coordinated cropping and dicing. Nat Rev Mol Cell Biol 6:376–385
Kiss T (2002) Small nucleolar RNAs: an abundant group of noncoding RNAs with diverse cellular functions. Cell 109:145–148
Kiss AM, Jády BE, Darzacq X et al. (2002) A Cajal body-specific pseudouridylation guide RNA is composed of two box H/ACA snoRNA-like domains. Nucleic Acids Res 30:4643–4649
Koonin EV, Senkevich TG, Dolja VV (2006) The ancient virus world and evolution of cells. Biol Dir 1:29
Levine MT, Jones CD, Kern AD et al. (2006) Novel genes derived from noncoding DNA in Drosophila melanogaster are frequently x-linked and exhibit testis-biased expression. Proc Natl Acad Sci U S A 103:9935–9939
Li J, Tang H, Mullen TM et al. (1999) A role for RNA helicase A in post-transcriptional regulation of HIV type 1. Proc Natl Acad Sci U S A 96:709–714
Lu L, Lee YRJL, Pan R et al. (2005) An internal motor kinesin is aassociated with the Golgi apparatus and plays a role in trichome morphogenesis in Arabidopsis. Mol Biol cell 16:811–823
Maizels A, Weiner AM (1999) The genomic tag hypothesis: modern viruses as molecular fossils of ancient stretegies for genomic replication. Biol Bull 196:327–330
Maizels N, Weiner AM, Yue D et al. (1999) New evidence for the genomic tag hypothesis: archaeal CCA-adding enzymes and tRNA substrates. Biol Bull 196:331–334

Makarova KS, Grishin NV, Koonin EV (2006) The HicAB cassette, a putative novel, RNA-targeting toxin-antitoxin system in archaea and bacteria. Bioinform 22:2581–2584
Matera AG, Terns RM, Terns MP (2007) Non-coding RNAs: lessons from the small nuclear and small nucleolar RNAs. Nat Rev Mol Cell Biol 8:209–220
Mattick JS (2001) Non-coding RNAs: the architects of eukaryotic complexity. EMBO Rep 2:986–991
Mattick JS (2003) Challenging the dogma: the hidden layer of non-protein-coding RNAs in complex organisms. Bioessays 25:930–939
Mattick JS (2006) The underworld of RNA. Nat Genet 38:393
Mattick JS (2007) A new paradigm for developmental biology. J Exp Biol 210:1526–1547
Mattick JS, Gagen MJ (2001) The Evolution of controlled multitasked gene networks: the role of introns and other noncoding RNAs in the development of complex organisms. Mol Biol Evol 18:1611–1630
Mattick JS, Makunin IV (2006) Non-coding RNA. Hum Mol Genet 15:17–29
Mattick J, Mehler MF (2008) RNA editing, DNA recoding and the evolution of human cognition. Trends Neurosci 31:227–233
Mehler MF, Mattick JS (2007) Noncoding RNAs and RNA editing in brain development, functional diversification, and neurological disease. Physiol Rev 87:799–823
Miller RK, Heller KK, Frisen L et al. (1998) The kinesin-related proteins Kip2p and Kip3p, function differently in nuclear migration in yeast. Mol Biol Cell 9:2051–2068
Miller S, Krijnse-Locker J (2008) Modification of intracellular membrane structures for virus replication. Nat Rev Microbiol 6:363–374
Mingle LA, Okuhama NN, Shi J et al. (2005) Localization of all seven messenger RNAs for the actin-polymerazitation nucleator arp2/3 complex in the protrusions of fibroblasts. J Cell Sci 118:2425–2433
Moazed D (2009) Small RNAs in transcriptional gene silencing and genome defence. Nature 457:413–420
Narayanan A, Lukowiak A, Jády BE et al. (1999) Nucleolar localization signals of box H/ACA small nucleolar RNAs. EMBO J 18:5120–5130
Nisole S, Saib A (2004) Early steps of retrovirus replicative cycle. Retrovirol 1:9
Nosek J, Kosa P, Tomaska L (2006) On the origin of telomeres: a glimpse at the pre-telomerase world. Bioessays 28:182–190
Obbard DJ, Gordon KHJ, Buck AH et al. (2009) The evolution of RNAi as a defence against viruses and transposable elements. Phil Trans R Soc B 364:99–115
Peterson-Burch BD, Voytas DF (2002) Genes of the pseudoviridae (Ty1/copia Retrotransposons). Mol Biol Evol 19:1832–1845
Piguet V, Sattenaut Q (2004) Dangerous liaisons at the virological synapse. J Clin Invest 114:605
Piriyapongsa J, Jordan IK (2008) Dual coding of siRNAs and miRNAs by plant transposable elements. RNA 14:1–8
Platini M, Goldberg I, Lamond AI et al. (2002) Cajal body dynamics and association with chromatin are ATP dependent. Nat Cell Biol 4:502–508
Poole AM, Jeffares DC, Penny D (1998) The path from the RNA world. J Mol Evol 46:1–17
Poole AM, Logan DT (2005) Modern mRNA proofreading and repair: proof that the last universal common ancestor (LUCA) had an RNA genome? Mol Biol Evol 22:1444–1455
Poon DT, Chertova EN, Ott DE (2002) Human immunodeficiency virus type 1 preferentially encapsidates genomic RNAs that encode Pr55(Gag): functional linkage between translation and RNA packaging. Virol 293:368–378
Presutti C, Rosati J, Vincenti S et al. (2006) Non coding RNA and brain. BMC Neurosci 7:5
Preuss ML, Kovar DR, Lee YRJ et al. (2004) A plant-specific kinesin binds to actin microfilaments and interacts with cortical microtubules in cotton fibers. Plant Physiol 136:3945–3955
Quellet DL, Perron MP, Gobeil LA et al. (2006) Micro RNAs in gene regulation: when the smallest govern it all. J Biomed Biotech 69616:1–20
Rashkova S, Karam SE, Kellum R et al. (2002) Gag proteins of the two Drosophila telomeric retrotransposons are targeted to chromosome ends. J Cell Biol 159:397–402

Reddy TR, Xu W, Wong-Staal F (2000) General effect of Sam68 on rev/Rex regulated expression of complex retroviruses. Oncogene 19:4071–4074

Richard P, Darzacq X, Bertrand E et al. (2003) A common sequence motif determines the Cajal body-specific localization of box H/ACA scaRNAs. EMBO J 22:4283–4293

Richard P, Kiss AM, Darzacq X et al. (2006) Cotranscriptional recognition of human intronic box H/ACA snoRNAs occurs in a splicing-independent manner. Mol Cell Biol 26:2540–2549

Rustom A, Saffrich R, Markovic I et al. (2004) Nanotubular highways for intercellular organelle transport. Science 303:1007–1010

Ryan FP (2004) Human endogenous retroviruses in health and disease: a symbiotic perspective. J R Soc Med 97:560–565

Ryan FP (2006). Genomic creativity and natural selection: a modern synthesis. Biol J Linn Soc 88:655–672

Ryan FP (2007) Viruses as symbionts. Symbiosis 44:11–21

Saunders WS, Koshland D, Eshel D et al. (1995) Saccharomyces cerevisieae kinesin- and dynein-related proteins required for anaphase chromosome segregation. J Cell Biol 128:617–624

Sfakianos JN, Hunter E (2003) M-PMV capsid transport is mediated by Env/Gag interactions at the pericentriolar recycling endosome. Traffic 4:671–680

Shadan FF, Villarreal LP (1996) The evolution of small DNA viruses of eukaryotes: past and present considerations. Viral Genes 11:239–257

Shav-Tal Y, Singer RH (2005) RNA localisation. J Cell Sci 118:4077–4081

Siomi H, Siomi MC (2009) On the road to reading the RNA-interference code. Nature 457: 396–404

Sontheimer EJ, Carthew RW (2005) Silence from within: endogenous siRNAs and miRNAs. Cell 122:9–12

St.Laurent G III, Wahlestedt C (2007) Noncoding RNAs: couplers of analog and digital information in nervous system function. Trends Neurosci 30:612–621

Sugiyama T, Cam H, Verdel A et al. (2005) RNA-dependent RNA polymerase is an essential component of a self-enforcing loop coupling heterochromatin assembly to siRNA production. Proc Natl Acad Sci 102:152–157

Takemura M (2001) Poxviruses and the origin of the eukaryotic nucleus. J Mol Evol 52:419–425

Tang Y, Winkler U, Fredd EO et al. (1999a) Cellular motor protein KIF-4 associates with retroviral Gag. J Virol 73:10508–10513

Tang PZ, Tsai-Morris CH, Dufau ML (1999b) A novel gonadotropin-regulated testicular RNA helicase. A new member of the dead-box family. J Biol Chem 274:37932–37940

Temin HM (1985) Reverse transcription in the eukaryotic genome: retroviruses, pararetroviruses, retrotransposons and retrotranscripts. Mol Biol Evol 2:455–468

Terada S, Hirokawa N (2000) Moving on to the cargo problem of microtubule-dependent motors in neurons. Curr Opin Neurobiol 10:566–573

Tomlinson RL, Ziegler TA, Supakorndej T et al. (2006) Cell cycle-regulated trafficking of human telomerase to telomeres. Mol Biol Cell 17:955–965

Tran E, Brown J, Maxwell ES (2004) Evolutionary origins of the RNA-guided nucleotide modification complexes: from the primitive translation apparatus? Trends Biochem Sci 29:343–350

Vale RD (2003) The molecular motor toolbox for intracellular transport. Cell 112:467–480

Van Lent JWM, Schmitt-Keichinger C (2006) Viral movement proteins induce tubule formation in plant and insect cells. In: Baluska F, Volmann D, Barlow P (eds) Cell-Cell Channels. Eurekah.com

Vetsigian K, Woese C, Goldenfeld N (2006) Collective evolution and the genetic code. Proc Natl Acad Sci U S A 103:10696–10701

Villarreal LP (2005) Viruses and the Evolution of Life. ASM Press, Washington

Villarreal LP (2009) Origin of Group Identity: Viruses, Addiction and Cooperation. Springer, New York

Villasante A, Abad JP, Planello R et al. (2007) Drosophila telomeric retrotransposons derived from an ancestral element that was recruited to replace telomerase. Gen Res 17:1909–1918

Wang C, Meier UT (2004) Architecture and assembly of mammalian H/ACA small nucleolar and telomerase ribonucleoproteins. EMBO J 23:1857–1867
Watanabe T, Totoki Y, Toyoda A et al. (2008) Endogenous siRNAs from naturally formed dsRNAs regulate transcripts in mouse oocytes. Nature 453:539–543
Weber MJ (2006) Mammalian small nucleolar RNAs are mobile genetic elements. PLoS Genet 2: e205
Witzany G (2006) Natural genome editing competences of viruses. Acta Biotheor 54:235–253
Witzany G (2007a) The agents of genomic creativity. In: Witzany G (ed) Biosemiotics in Transdisciplinary Contexts. Proceedings of the Gathering 6 in Biosemiotics. Umweb, Helsinki
Witzany G (2007b) The *Logos* of the *Bios* 2. Bio-Copmmunication. Umweb, Helsinki
Witzany G (2008a) Bio-Communication of bacteria and their evolutionary roots in natural genome editing competences of viruses. Open Evol J 2:44–54
Witzany G (2008b) The viral origins of telomeres, telomerases and their important role in eukaryogenesis and genome maintenance. Biosemiotics 2:191–206
Yazgan O, Krebs JE (2007) Noncoding but nonexpendable: transcriptional regulation by large noncoding RNA in eukaryotes. Biochem Cell Biol 85:484–496
Yedavalli VS, Neuveut C, Chi YH et al. (2004) Requirement of DDX3 dead box RNA helicase for HIV – I Rev-RRE export function. Cell 119:381–392
Zemann A, Beckke A, Kiefmann M et al. (2006) Evolution of small nucleolar RNAs in nematodes. Nucleic Acids Res 34:2676–2685

Chapter 11
Outlook

11.1 From Mechanistic Biology to Biocommunication

In the second half of the twentieth century the physiological processes of all levels of cells, tissues, organs and organisms of all organismic kingdoms were the mainstream focus of biological research and experiments. In the 1970s, increasing use of 'communication'-metaphor occurred. In the last decade of the twentieth century, communication (no longer used as metaphor) within and between organisms surpassed the pure physiological understanding of organisms. Cell-cell-communication dominated contemporary cell biology, including considerable knowledge about a great variety of signalling pathways serving as both organisation and coordination of production, release, uptake and information-processing within and between cells.[1]

[1] The Editor of *Cell Communication and Signaling*, Bernard Perbal, writes in his editorial 'Communication is the key' (2003): 'All forms of communication between human beings have long been recognized as a requirement for reciprocal understanding, transfer of knowledge, and productive development of societies. This also applies to living cells who are organized in «microsocieties» that constantly adjust to their environment through a complex network of signaling pathways. The chemical communication which occurs at various levels results in an integrated exchange of information that is essential for coordinated responses'.

Research on biocommunication including all levels of biological organisation is also the goal of a new journal *Communicative & Integrative Biology* (2007): In 'about' we read: 'Entering the new millennium – equipped with all sorts of -omics data growing at an incredible speed – the biological sciences are slowly moving from a reductionistic to a holistic view of biological complexity. One such example is the complex organismal behavior based on signaling and communication processes. The extreme intricacy of these processes requires integrated systemic approaches. This new journal, *Communicative & Integrative Biology*, will serve as a platform for the synthesis of the biological sciences. The focus will be organismal communication, however, *Communicative & Integrative Biology* will deal with communication at all levels of biological organization from subcellular organelles to societies, ecosystems and the biosphere as a whole'.

Additionally the scope of an online journal published by *Science* (2008) has to be mentioned: '*Science Signaling* will publish leading research papers related to the broad topic of signal transduction. Appropriate topics include any analysis of the mechanisms by which cellular functions are regulated in response to external or internal cues in both eukaryotic and prokaryotic cells. The subject matter will thus cross the boundaries of traditional research areas such as Cancer Biology, Cell Cycle Regulation, Cell Biology, Biochemistry, Development, Microbiology, Molecular Biology, Immunology, Neuroscience, Physiology and Medicine, Pharmacology, and Plant Biology. Papers

G. Witzany, *Biocommunication and Natural Genome Editing*,
DOI 10.1007/978-90-481-3319-2_11, © Springer Science+Business Media B.V. 2010

The link between linguistics and genetics has been obvious since the detection of the universal grammar and the structural code of DNA. Chomsky's meaning-independent syntactical approach (i.e. the primacy of logics of the syntactic structure) led to the broad acceptance and usage of bioinformatic methods and systems biology. Additionally, researchers in bacteria communication like Ben Jacob suggested with good reason that this approach reduces linguistic competences found in bacterial communication and has to be satisfied by both semantic aspects, i.e. the context-dependent meaning of signals which act as signs, and pragmatic aspects, *which focus on the variety and differences of the behavioural patterns in common-goal coordination, shared knowledge, memory and mutual intentions.* Apart from that, it is coherent with the presupposition by Charles Morris of any non-reductionistic analysis of language-like structures, the complementarity of syntax (combinatorial rules), pragmatics (agent dependent contextual rules) and semantics (content-specific rules).

This suggestion is also valid in genetics and genomics. Also here the pure molecular syntax of the sequences has to be satisfied by both semantic aspects, i.e. the context-dependent meaning of sequences, and pragmatic aspects which focus on the variety and differences of the behavioural patterns *of an organismic identity* embedded in species-specific and, parallel with this, in trans-specific interactions. Participants on the conference in 2001 'Contextualizing the Genome: the Role of Epigenetics in Genetics, Development and Evolution' exemplified in great detail that this pragmatic context is indispensable for a non-reductionistic understanding of epigenetics.

As in the theoretical investigation of human language and communication, all attempts to get a coherent meaning of content exclusively from an analysis of the syntactic structure of characters failed. The message alone is nothing; its meaning depends on both the situational context in which it is used by a *competent agent* and on the expressional patterns used by this agent which determine the meaning that should be transported. *The identical form of a syntactical structure therefore can transport contradictory messages.* This is a crucial deficiency of algorithmic analyses in the realm of information-theoretical approaches or mathematical theories of biology.

In parallel, the usage of 'language'-metaphor has increased since the mid-twentieth century with the growing knowledge about this genetic code. Most of the processes which evolve, constitute, conserve, rearrange the genetic storage medium DNA are terms which were originally used in linguistics such as coding, copying, transcription, translation, signalling, signal transduction, etc. Meanwhile the linguistic approach has also lost its metaphorical character and the similarity between linguistic languages/codes and the genetic storage medium are not only accepted but are fully adapted in bioinformatics, biolinguistics, protein linguistics,

---

should substantially refine current understanding of important signaling processes with priority given to those papers that provide new concepts and new understanding of biological signal transduction and that are likely to find application in multiple biological systems or in a diverse range of investigations'.

biohermeneutics and biosemiotics. The advantage of methodological adaptation of communication and linguistic terminology is the availability of appropriate tools for differentiation at specific levels which are difficult to describe in the language of physics and chemistry alone.

This results in language-like structures and communication processes occurring on the back of living nature which are not at all evolutionary inventions of humans, nor are they anthropomorphous adaptations of terms of description of human capabilities on non-human living nature. It was and is obvious that every coordination and organisation within and between cells, tissues, organs and organisms *needs signs*, i.e. chemical molecules which serve as signals or symbols in messages or serve as vital indicators of environmental conditions which can be interpreted by organisms and therefore serve as important indices. These signs need to be interpreted in a correct way by biological agents, i.e. there must be subjects/representatives of sign production and sign interpretation. This means that the interpretation may also fail and result in the lack of or inappropriate response behaviour.

Biocommunicative acts are the precondition for coordination within and between organisms to take place. Biocommunication occurs on three levels: (A) intra-organismic, i.e. intra- and intercellular, (B) inter-organismic, between the same or related species and (C) trans-organismic, between organisms which are not related and in most cases are of different organismic kingdoms, as happens in most symbioses such as lichens.

The pragmatic philosophy of biology which I developed between 1987 and 1990 supposes that, on the assumption that life is structured linguistically and organised communicatively, there must be a complementarity of three levels of rules wherever signs are used within (inner nature) and between (outer nature) living organisms. Because one sign is not a sign, and one sign-user cannot use signs without a community of sign-users which he/she/it is a part of (whether by socialisation, i.e. memory and learning, or by innate competence transfer), there is a close relationship to same or related species which share (A) syntactic, (B) pragmatic and (C) semantic rules of sign use, i.e. (a) how signs can be combined or sequenced (syntactic rules), (b) how interactions like coordination and organisation can be established (pragmatic rules) for different real-life purposes like common behaviour, or attack, defence, mating, virulence, nutrition uptake, etc. Dependent on the *pragmatic context* are the semantic rules which determine the meaning of signs, i.e. (c) how one syntactic sequence can have completely different meanings in different situational contexts.

11.2 Three Kinds of Signs in Biocommunication

According to the founder of semiotics, Charles Sanders Peirce, we are able to differentiate three kinds of signs: Indices, Icons, Symbols. Within the current biocommunicative approach *Indices* are in most cases abiotic stimuli from the environment which are *interpreted* as signals, e.g. a plant root identifies nutrients as being relevant, as the plant shoot does with the angle of sunlight. *Icons* are biotic one-to-one signals (*analogue*) which need no further explanation, e.g. plant cells identify auxin within a hormonal coordination process, and there are *symbols*, i.e. signs or sequences of signs like characters of an alphabet used to generate words, sentences, codes which do not indicate by themselves what they mean (what their function could be) but are signs through natural or cultural *conventions*. Such sequences may also be sequences of behaviours such as the dances of the honey-bees of the colder hemisphere or even the eukaryotic genome-grammar with its high complexity of alternately coding and non-coding sections.

11.3 Context Determines Meaning

Because chemical molecules which serve as signs are not indefinite it is normal for sign-using organisms to share a finite number of signs and a finite number of semiotic rules which are necessary for use, so it is common for living organisms to use the same chemical substances (in liquid, gaseous or mineralised form) which serve as signs to transport messages in different pragmatic (situational) contexts; auxin in plant communication may transport various messages whether it is used in the context of (i) neuronal-like cell-cell communication or (ii) as hormone to transport messages between root and shoot or (iii) as a morphogenic sign. Some dances of the bees serve as symbols to transport the message of an appropriate hive or in another pragmatic (situational) context to transport the message of an appropriate place to find nectar. In other cases, too, the specific geographical situation of the populations for growing up and socialisation may lead to special dialects, i.e. the signs which are used to communicate are identical but the meaning of the transported messages differs depending on regional customs and traditions. This phenomenon of dialects we find also in bacterial communication. Dialects in biocommunicative acts could be an important feature in the investigation of in vivo habitats.

11.4 Living Nature and Non-living Nature

The difference from pure stochastic/mechanistic biology is that semiotic rules of sign use cannot be explained sufficiently by the natural laws of physics and chemistry alone because, differently from pure natural laws such as gravity, living organisms are in continued relations with semiotic rules and do not have a changeable relation with natural laws but underlie them in a strict manner.

That means rules of sign use may be altered, rearranged, invented, generated or deleted, i.e. communication processes are undertaken to function but in every case they may also fail, i.e. there is no natural law which guarantees success. The success of a rule-governed sign-mediated process, e.g. communication, is not the result of the previous status of all circumstances. The following of semiotic rules in sign-use and -interpretation is different from e.g. the crystallisation of water to ice. This is also the case with the genetic code. If the genetic code functions as code this is physically and chemically coherent. The biotic plus is that this code also functions syntactically, pragmatically and semantically in parallel, which the freezing of water to ice does not. There is no semiotic interpretation of water molecules which are exposed to some degrees under zero apart from physical and chemical laws alone.

Each of these levels of semiotic rules may be damaged or deformed by several influences and circumstances, so that the content of the code is not correct, i.e. is not able to serve as matrix for appropriate protein production (coding) or as element of higher-order regulation (non-coding).

These differences mark the border between non-living matter and living nature. Therefore it becomes clear that the conditions on the surface of a planet without living organisms are determined entirely by the sun's energy and the laws of physics and chemistry, as James Lovelock showed convincingly. A planet with living organisms deviates considerably from this scheme. The gas composition and temperature will vary in a manner that cannot be predicted exclusively by the laws of physics and chemistry alone. Such a planet contains incompatible gas mixtures and temperatures whose relatively stable balance is actively controlled by organisms, and these organisms are no *solus ipse* subjects but part of a planetary symbiology which is organised and coordinated by continued biocommunicative acts.

11.5 Biocommunication Defines a Biotic 'Plus'

If we choose the biocommunicative approach we can steer towards a new, transparent and, for every living organism, coherent method of description of the levels of interaction in which the organism is integrated. The non-human living world does not exclude humans, as we are part of living nature with indispensable interrelations. The old system that categorises humans on one side, the environmental living beings on the other, is not appropriate nowadays. Without symbioses there would not have been something like living nature in principle, and beneath intraspecific and species-specific sign-use there is also a level of trans-species sign use.

It is clear that the 'biotic plus' of living nature and therefore the difference between non-living and living nature started in very few or only one initial event. The starting-point of living nature is the use of sign(al)s between agents competent to use, produce and interpret signs according to syntactic, pragmatic and semantic rules. This starting-point could also fail and turn back into the physico-chemical realm without semiotic rules again. The start of living nature from a biocommunicative perspective may have had several processes of trial and error, but in contrast to replicative copying of nucleic acids via chemical basepairing it must be stated clearly that from the biocommunicative approach additionally (en)coding competences are the key features of starting life. Coding competences without replicative copying would not function but copying without coding can't be part of a definition of life.

11.6 The Advantages of Biocommunicative Biology

Therefore, is the aim of a biocommunicative approach to observe and describe the semiotic rules of sign use in living nature? If the description is correct, our understanding of coordination and organisation of evolution, development, growth and disease in the living nature (cells, tissues, organs, organisms) will increase. This could be an advantage in medical sciences, as proven by the work of Thure von Uexküll.

In some cases we will be able to predict vectors of development but the most important advantage of this new type of biology (as a kind of an, understanding' social science)[2] over the former mechanistic biology is that we can be aware of the complexity of living nature to prevent careless or even negligent steps in technical developments, e.g. the use of artificial bio-engineering without being aware of possible consequences. There is a crucial difference whether we try to handle a mechanical, functional matter or communicatively organised living nature. In the first case we can trust laws of physics and chemistry exclusively; in the second we must be aware that change and reorganisation, success and failure, of biotic agents are common and are an open process in principle.

11.7 Linguistic and Communicative Competences in Non-human Nature

The three semiotic levels of rules which are necessary to generate a shared appropriate repertoire for sign-mediated interactions, i.e. communication processes, have a complementary function. Methodologically this means that if one level is reduced to another (or the remaining others) no adequate analysis of sign-use is possible: through analysis of syntax or semantics alone it is not possible to understand or even reconstruct the pragmatic interactional content of two communicating agents, and vice versa. This is a crucial difference between a biocommunication approach and e.g. bioinformatics, with its main focus on genetic syntax, or biosemantics, with its main focus on meaning functions, or even biosemiotics, which primarily investigates signs and sign-processes, not *interactions* of organisms.

There is also a difference in (A) constructing correct sentences or genetic sequences according to syntactic, pragmatic and semantic rules and (B) generating an appropriate behaviour in which the sequence of behaviour, e.g. attack, defence, mating, virulence, common coordination for a special purpose, takes the role of signs. The competence to invent or rearrange the genomic content is a kind of

[2] At the current stage biology is an empirical science which explains biological observations. In the philosophy of science discourse in the last 1970s and 1980s there was an 'explaining-understanding' controversy (between logical empiricism and hermeneutics). The concept of 'biology as an understanding social science' would be a progress out of the results of this discourse.

linguistic competence which includes other semiotic rules apart from the *communicative competence* to initiate and constitute a context-coherent *interaction* in all its steps und accompanying sub-steps. Therefore biocommunication of genome editing includes other issues apart from biocommunication of e.g. plants, bacteria, animals or fungi, although these two levels of biocommunication are connected in natural genetic and even more in natural epigenetic engineering.

11.8 Complementary Roles of Linguistic and Communicative Competences

As we could see, the linguistic competence of viruses (and especially their competence to constitute the sequence ('word') order in bacteria) leads to all necessary prerequisites for the genetic and genomic as well as proteomic evolution. This viral linguistic competence needs living organisms which are different to each other, however, needs a biotic 'matrix' to expand these competences. Without living and interacting bodies, cells, etc. the genomic creativity would be only a possibility restricted to mere RNA combinatorial events (in an early pre-cellular RNA world) without being relevant to the generation of a biosphere. Agents with genome editing competences need a real lifeworld of interacting metabolising cell bodies to get different situations as matrix for inventions or genomic creativity, as do organisms which cannot live without continued natural genetic engineering and genome editing within their cells. This means linguistic competences need the complementary role of communicative competences.

11.9 New Qualities for Future Decisions

The origin of life is not questioned in this book, which is focused primarily on communication processes, i.e., rule-governed, sign-mediated interactions within and between living organisms on all levels in all organismic kingdoms. Furthermore, it develops these for an appropriate description of the linguistic competences of agents which are editing the genetic/genomic contents.

It should provide a better understanding of living nature in general and what it means to be part of this ourselves. Differently from pure knowledge-shuffling for commercial purposes, this biocommunicative approach may contribute to change the relationship of human beings with non-human living nature. This non-human living nature is not the environment, the *Umwelt*, but in a symbiotic living nature a *Mitwelt*, a co-world on which we vitally depend. If our ecological 'footprints' are destructive further down the line we counteract symbiological rules of living nature. This could mark a copernicanian turn in our relation and understanding of life in general.

It is our decision which pragmatic directions we want to go in, a destructive, parasitic cycle of exploitation, pollution and extinction of species or a responsible sustainable and symbiotic-like cooperation with living nature to human and non-human benefit. The success of the latter will decide in the long run whether our understanding that we are at the top of the evolutionary ladder is correct or, conversely, whether humankind is an evolutionary error. As living organisms we have to decide (not as *solus ipse* subjects but by majority vote) to communicate, to coordinate and to organise such practical behaviour either in a sustainable coherence to living nature or as overdeveloped hybrids of the species anthropos.

Index

Note: The locators following with the letter 'f' in the index denotes figures.

A

Abiotic, 24, 28, 30, 45f, 70, 74, 92–93, 102, 103f, 110, 111, 123f, 188, 199
Active agents, 160
Adaptational purposes, 24, 110, 123, 173
Adaptive immune system, 129, 139, 140, 141, 142, 191
Addiction modules, 120, 134, 135, 139, 140, 141, 142, 161, 162, 176–180, 183, 184, 187, 190, 191
Agents, 19, 20, 21, 22, 23, 24, 45f, 61, 76, 80f, 96, 119, 121, 129–145, 151, 152, 153, 159, 160, 161, 162, 163–164, 166, 172, 173, 174, 175–180, 182, 183, 184, 185, 187, 188–190, 198, 199, 201, 202, 203
Algorithms, 11, 13, 14, 21, 62, 198
Alternative splicing, 21, 42, 130, 160, 166, 178, 179, 180
ALUs, 142, 184
Antitoxin, 101, 120, 134, 141, 176, 191
Archaea, 24, 71, 78f, 80f, 110, 118, 119, 122, 137, 144, 180, 185, 186, 187, 189
Archaeal phage, 138, 153, 165, 174, 175
Archaic viral strategy, 134
Artificial genetic engineering, 131
Artificial language, 10, 12, 13
Atomism, 5, 16
Auto-silencing, 181
Auxin, 29, 30f, 41, 200

B

Background memory, 123
Bacterial community, 110, 111, 115, 116, 122, 123
Bee colonies, 54, 55, 57, 59, 62, 64
Bee language, 57, 59–60, 63
Behavioural context, 33, 45, 46, 61, 98, 99, 100, 102
Behavioural patterns, 23, 42, 43, 54, 68, 69, 75, 90, 91, 93, 96, 97, 103, 110, 111, 122, 123, 198
Beneficial, 28, 32, 68, 72, 81, 93, 94, 95, 100, 101, 102, 110, 113, 114f, 115, 191
Binary codes, 9, 10, 16
Biocommunication, 20, 23–24, 42, 54, 67–82, 89–103, 197–200, 201, 202, 203
Biocommunication of bacteria, 109–123
Biofilm organization, 110, 111, 112, 116f
Biohermeneutics, 22, 199
Bioinformatics, 21–22, 111, 198, 202
Biolinguistics, 21–22, 198
Biosemiotics, 22, 44, 46, 158, 199, 202
Biotic matrix, 121–122
Biotic signals, 37, 102

C

Cajal bodies, 161, 185, 186, 189
Capsid, 120, 121, 122, 175, 179
Caribbean reefs, 68
Cell biology, 172, 197
Cell-cell communication, 28, 29, 38, 41, 116, 197, 200
Cell–cell recognition, 117
Cell surface, 112, 121
Cellular network, 101
Cellular responses, 97, 99
Cell wall structures, 32
Centromeres, 150, 157, 160, 165, 166, 181
Chance mutations, 62, 118, 144
Characteristica universalis, 16
Chemical communication, 28, 29, 197
Chromatin markers, 130
Chromosomal rearrangements, 41

G. Witzany, *Biocommunication and Natural Genome Editing*,
DOI 10.1007/978-90-481-3319-2,

Chromovirus, 101, 102
Circular genomes, 120, 153, 154
Circular ssRNAs, 152
Cnidaria, 79
Coding, 12, 41, 42, 100, 129, 130, 131, 137, 143, 150–151, 153, 163, 164, 165, 166, 198, 200, 201
Coding-fidelity, 153
Coding sequences, 129, 130, 131, 132, 142, 160, 164, 166, 172
Combinatorial communication, 99
Communication
  patterns, 68, 69, 74, 102, 117, 123f
  rules, 20
  sciences, 18
  theory, 19
Communicative actions, 19, 20
Communicative competence, 19, 20, 24, 25, 28, 34, 44, 46, 53–64, 68, 109–111, 113, 119, 123, 152, 174, 202–203
Communicative competences of plants, 24, 28, 46
Communicative complexity, 110
Communicative intersubjectivity, 18–20
Communicative living nature, 20
Communicative practice, 18
Communicative rationality, 20
Conjugation, 121
Contact-specific signalling, 111
Content-specific, 24, 163, 172, 190, 198
Context, 12, 14, 19, 21, 22, 23, 24, 28, 29, 31, 33, 42, 43, 45, 46, 60–64, 68, 69, 70, 73, 81, 96f, 98, 99, 100, 102, 111, 115, 116, 118, 123, 140, 143, 144, 163, 165, 166, 172, 173, 176, 177, 179, 185, 187, 188, 190, 198, 199, 200, 203
Context-dependent, 29, 68, 111, 115, 123, 173, 198
Context-sensitive, 24, 172, 190
Contextualization, 118
Coordinate, 17, 20, 23, 30, 32, 33, 37, 39, 53, 60, 62, 68, 72, 74, 77, 90, 91, 92, 97, 98, 99, 103, 109, 110, 111, 112, 113, 115, 131, 136, 138, 150, 163, 165, 172, 173, 180, 182, 185, 201
Coordination of common behavior, 152
Coordination processes, 53, 200
Copia, 160
Copying, 131, 150–151, 159, 188, 198, 201
Coral animals, 24, 68, 70, 72, 79, 80, 120
Coral disease, 68, 120
Coral reefs, 67, 80
Co-suppression, 130
Co-world, 203
Critical rationalism, 7, 15
Cross-kingdom signalling, 116
Cross-talk, 91, 98
Cyanophage, 134, 138, 174
Cybernetic system theory, 9

D

DEAD box proteins, 179
Decentral, 29, 37, 43, 79
Decision-making, 43, 62, 123
Decoding, 12, 15, 18, 79
Deep grammar, 164–165
Defence strategies, 32, 33
de novo generation, 189
Depiction, 2, 8, 9, 10, 11–13, 15, 16, 21, 23
Determinism, 5
Development, 2, 3, 5, 6, 7, 11, 22, 27, 28, 29, 30, 31, 32, 35, 36, 37, 38, 39, 44, 46, 54, 68, 77, 79, 80, 81, 91, 93, 97, 98, 103, 111, 115, 118, 130, 140, 141, 143, 173, 178, 182, 186, 198, 202
Developmental processes, 11, 14, 19, 39, 41, 43, 110, 112, 144, 177, 180
Dialects, 3, 4, 59–60, 64, 116f, 200
Dicer, 176
DNA-editing processes, 130
DNA invention, 188
DNA language, 164
DNA methylation, 41, 143, 183
DNA movements, 121
DNA rearrangement, 130, 160, 162, 191
DNA repair pathway, 157
DNA storage, 39, 99f, 131, 150, 152, 153, 163, 171, 181, 189, 190
DNA-textbook conviction, 42
DNA viruses, 175, 176, 177, 178, 179, 189
Double-stranded DNA viruses, 100, 120, 132, 133, 165
Double-stranded RNA viruses, 100, 102, 133, 161
Dynein, 178–180

E

Electrical signals, 29, 34
Elementary sentences, 15
Embryogenesis, 38
Emergentism, 3
Empirical data, 20, 118
Encoding, 15, 18, 138, 181
Endocytosis, 39, 41, 42, 178
Endogenous agents, 163, 164
Endogenous retroviral swarms, 131, 142

Endogenous retrovirus, 101, 102, 120, 132, 133, 140, 142, 143, 150, 163, 177, 183
Endogenous RNA species, 181
Endophytic fungi, 35, 94
Endosymbiotic partner, 120
End replication problem, 158
Endsoymbionts, 78
Epigenetic, 21, 31, 39, 40f, 41, 42, 61, 62, 78f, 111, 130, 131, 144, 189, 198, 203
Epigenetic regulation, 159, 180, 181
ERV, 133, 143
Eubacterial phage, 138, 174
Eukarya, 71, 137, 162, 184, 186, 187
Eukarya-like dsDNA viruses, 152
Eukaryotic cell, 39, 40, 41, 77, 136, 149, 159, 162, 163, 165, 174, 176, 178, 179, 185
Eukaryotic nucleus, 40, 121, 122, 129, 136–139, 149, 153, 162, 165, 166, 174–175, 177, 178, 189
Eukaryotic superkingdom, 149
Everyday language, 12, 15, 16, 17, 18, 130
Evolution, 7, 11, 13, 21, 24, 31, 35, 40, 44, 46, 67, 79, 80, 81, 89, 100, 102, 103, 110, 115, 117, 118, 119, 121, 122, 129, 130, 133, 134, 140, 141, 142–143, 144, 149, 151–152, 153, 154, 159, 160, 162, 164, 166, 174, 175, 178, 186, 187, 188, 198, 202, 203
Evolutionary history, 20, 93, 102, 119, 132, 144, 149, 154, 165
Evolutionary modifications, 129
Evolutionary protocols, 171, 181, 188, 190
Exact sciences, 7, 8, 9, 16, 23
Excision, 41, 131, 152, 184
Excision of introns, 152
Expression factors, 133

F
Feedback, 32
Feedforward, 32
Filamental tips, 90
Filamentous fungi, 101, 158
Floral patterning, 31
Flowering plants, 24, 94, 139
Formalisable, 8, 9, 10, 12, 16, 17, 18, 21, 23
Formalised language, 8
Formal language, 13, 14
Formal system, 11, 13, 14, 21
Fruiting bodies, 43, 91, 112, 113, 116f, 123
Fungal diseases, 89, 91, 95
Fungal mats, 94, 101
Fungal networks, 101
Fungal spores, 89, 93
Fungi, 24, 32, 34–35, 35f, 36, 41, 42, 45f, 71f, 72, 73, 74, 78f, 80f, 89, 90, 91–92, 93, 94, 95, 96, 96f, 97, 100–102, 110, 113, 114f, 116, 133, 158, 161, 176, 203

G
Gag protein, 159
GenBank database, 137, 143
Gene-blocks, 117, 118
Gene expression, 31, 41, 75, 79, 96, 110, 115, 142, 180–188, 191
Gene order, 110, 117
Gene pool, 117, 118, 152, 164
Gene silencing, 130, 161
Genetic alphabet, 172
Genetic alterations, 130
Genetic arrangements, 121, 171
Genetic code, 9, 21, 42, 144, 152, 172, 173, 174, 198, 201
Genetic complexity, 130
Genetic data set, 21, 42, 111, 118, 119, 122, 130, 144, 152, 166, 189
Genetic exchange, 117, 118
Genetic expression, 78, 112, 172, 183
Genetic flux, 122
Genetic identity, 120, 133, 176
Genetic parasites, 100, 142, 150, 153, 159, 160, 162, 163, 165, 175, 176–180, 182
Genetic settler, 101, 120, 153, 160, 173, 178, 190
Genetic text-editing, 151
Genome
architecture, 42, 129, 131
-editing, 166
-editing competences, 24, 118f, 122, 129–145, 163, 165, 174, 175, 187, 203
Genomic creativity, 136, 163, 185, 203
Genomic identity, 120, 190
Genomic innovation, 118, 136, 144, 190
Genomic plasticity, 43, 122
Group I-introns, 137
Group II self-splicing introns, 161
Gypsy, 160

H
H/ACA RNAs, 185
HERV, 143, 160
HetA, 159
Heterochromatin formation, 157, 181
Higher-order regulations, 42, 130, 164, 201
Histone modifications, 41, 42, 130, 180
History of sciences, 19
Hive, 54, 55, 56, 57, 59, 61, 62, 63, 200

Holism, 3, 4
Honey-bee, 20, 53–64
Horizontal gene transfer, 117, 118, 118f, 121, 122, 139, 153, 164
Hormone-like signalling, 114
Hormones, 29–30, 31, 35, 38, 69, 69f, 77, 92f, 94, 112f, 114, 200
Host genomes, 24, 39, 101, 102, 119, 120, 121, 122, 133, 142, 159, 172, 173, 175, 176, 177, 178, 182, 183, 187, 191
Host signal recognition, 117
Human oral cavity, 117
Hylozoism, 3
Hyphal growth, 90, 92, 97, 113

I
IAPs, 160
Icon, 199, 200
Immune response, 95, 139, 140, 172, 177, 191
Immune system, 101, 120, 129, 133, 134, 139, 140, 141, 142, 143, 178, 190, 191
Immunological synapse, 34
Immuno-reactive response, 142
Imprinting, 21, 61, 130
Incompleteness theorem, 13–14
Index(ices), 54, 92–93, 199
Inflammation, 140
Information theory, 9, 10, 12, 22, 23
Inheritable datasets, 131
Innate immune system, 101, 133, 139, 140
Innovation-sharing protocol, 144
Inorganic matter, 7, 19
Insects, 29, 30, 32, 33, 34, 35f, 36, 37, 42, 45f, 46, 93, 94, 94f, 95, 103f, 114, 134, 158
Insertion, 41, 131, 176, 184, 191
Integrase, 132, 138, 141, 177, 179
Interactional context, 102, 103, 111, 143
Interbacterial communication, 114
Intercellular communication, 37–39, 77, 95, 97–98, 114
Intermediate storage, 131
Internal communication, 68, 92
Inter-organismic communication, 24, 54, 95
Interpretation processes, 12, 22, 24, 32, 53
Interpreter, 28
Interspecific signalling, 115
Intersubjective interactions, 18
Intracellular communication, 25, 31, 39–42, 77–79, 98–100, 102, 118–119
Intracellular signals, 29, 99
Intra-organismic communication, 25, 77, 78f, 80, 92

J
Jitter-dance, 58
Junk DNA, 129, 130, 172

K
Key agents of DNA replication, 163
Killer toxins, 102
Killer viruses, 100, 101
Kinesin, 178–180
Knowledge, 4, 5, 6, 8, 12, 18, 19, 20, 22, 23, 24, 111, 172, 182, 190, 197, 198, 203

L
Languages
of bacteria, 110
games, 7, 12, 14, 16, 17, 18, 19
-like codes, 9, 20, 111, 198, 199
LaPlace demon, 5
Last universal common ancestors, 151, 152
Laws of physics, 144, 200, 201, 202
Learning, 11, 40, 42, 43, 60, 61, 64, 116f, 181, 199
Lichens, 90, 93, 95, 199
Life sciences, 20
Lifeworld, 2, 11, 12, 16, 19, 64, 203
Linear chromosomes, 136, 138, 139, 152, 153, 154, 157, 158, 165, 174, 175, 189
LINEs, 101, 130, 142, 143, 160, 161, 164, 184, 190
Linguistic action, 53
Linguistic competence, 12, 18, 58, 111, 152, 174, 198, 203
Linguistics, 21, 23, 111, 172, 198
Linguistic sign, 10, 12, 15, 53, 54, 55, 57, 58, 59, 60, 61, 62, 63, 64
Linguistic turn, 8–11, 16–18, 19
Living nature, 1, 14, 20, 22, 28, 44, 120, 150, 199, 200–201, 202, 203, 204
Logical empiricism, 7, 8, 15, 202
Logical structure, 9, 10, 11, 14, 15, 16, 17
Logics, 2, 10, 15, 198
Logic of science, 1
Long interspersed elements, 130
Long terminal repeats, 130, 164
Long-time storage data, 131
LTRs, 130, 142, 159, 160, 161, 164, 190
Lytic, 79, 96, 119–121, 132, 133, 135, 139, 144, 176, 177, 178, 191

M
Materialism, 4, 8, 22
Mathematical language theory, 10
Mathematical reductionism, 18
Mathematics, 9, 10, 13, 14, 15, 16, 17

Mature mRNA, 163
Meaning, 9, 15, 16, 17, 18, 19, 20, 21, 22, 23, 28, 42, 43, 46, 58, 59, 60–64, 68, 81, 100, 102, 103, 111, 114, 123, 130, 171, 172, 188, 190, 198, 199, 200, 202
Meaningfully recombining, 135
Meaning-independent syntax, 9, 15, 17, 23, 43, 58–59, 60–64, 68, 130, 131, 144, 164, 172, 189, 199, 200
Mechanical signs, 28
Mechanicism, 20, 22
Memory, 27, 32, 36f, 42, 43, 57, 61, 75f, 79, 96f, 111, 115, 116f, 123, 180, 181, 198, 199
Messenger RNA, 172, 181, 182
Messenger substances, 27, 29, 31, 34, 38
Metahormones, 30
Meta-language, 13, 14, 16, 18
Metaphysics, 4, 7–14, 16
Microbes, 29, 32, 33, 35, 37, 70, 71, 72
Micro-RNAs, 31, 36, 41, 42, 130, 133, 137, 163, 176
Mitochondrial plasmids, 161
Mitwelt, 203
Mobile agents, 163, 174
Mobile elements, 159–160, 165, 184, 186
Modification, 41, 42, 101, 121, 122, 129, 130, 131, 133, 134, 135, 138, 139, 144, 162, 166, 175, 176, 180, 184, 185, 186, 189, 190, 191
Molecular languages, 10, 28
Molecular syntax, 21, 28, 81, 120, 123, 143, 144, 150, 158–159, 165, 166, 172, 181, 198
Molecular vocabulary, 29
Monism, 3, 4
MRNA, 38, 152, 153, 160, 162, 163, 164, 175, 176, 179, 180, 181, 185, 191
MRNA degradation, 176, 181
MRNA translation inhibition, 176
Multi-cellular, 77, 95
Multiplicity reactivation, 134–135
Mushrooms, 89
Mutualism, 45, 110
Mutualistic bacteria, 117
Mycelia, 34, 90, 95
Mycobacterial phage, 138
Mycorrhizal fungi, 34, 35, 35f, 94, 113

N

Natural genetic engineering, 109, 131, 149, 151, 163, 166, 185, 189, 203
Natural genome editors, 24, 61, 118f, 119, 121–122, 123f, 129–145, 153, 160–162, 163–164, 165, 166, 173, 174, 175, 185, 187, 190
Natural laws, 5, 6, 9, 10, 11, 19, 44, 81, 200
Neighbouring plants, 33, 36, 37, 42
Nervous system, 7, 10, 11, 29, 37, 43, 143
Neuronal, 29, 34, 37, 40, 41, 42, 43, 44, 79, 115, 181, 182, 185, 200
Neuronal plasticity, 40, 41, 79
Neuron populations, 40
Nominalism, 8
Non-coding RNA populations, 163
Non-coding RNAs, 24, 112, 160, 164, 174, 180, 181, 182, 183–184, 189, 190
Non-coding sections, 131, 166, 200
Non-living nature, 1, 200–201
Non-LTR retroposons, 159, 161
Non-lytic, 24, 122, 131, 138, 151, 163, 166, 173, 175, 176, 178, 189
Non-lytic viral settlers, 176
Non-protein-coding underworld, 165
Non-random genetic change operators, 163
Non-reductionistic, 14, 20, 111, 198
Non-self, 33, 36, 36f, 39, 68, 71, 75, 75f, 76, 90, 95, 96f, 97, 99f, 113, 116f, 117, 122, 140, 151, 152, 160, 163, 174, 175, 181, 182, 191
Non-self agents, 45f, 141, 188
Novelty, 6, 7
Nuclear membrane, 136, 162, 178, 183
Nuclear pores, 136, 139, 152, 165, 174, 175, 178, 183
Nuclear RNA, 162, 163, 180, 181, 182, 186
Nucleic acid language, 11, 152, 172, 173, 174, 188, 191
Nucleolus, 161, 183, 185, 186
Nucleoprotein structures, 152, 157
Nucleotide grammar-editing abilities, 151
Nucleotide word, 120, 144, 151, 181
Nucleotide word order, 144, 151, 181
Nutrient-sensing signals, 99

O

Objectivism, 8, 19, 22
Objectivity, 8, 18
Object language, 13, 14
Observational languages, 13
Occidental philosophy, 1
Ontology, 7, 15, 22
ORF, 161
Organisation, 7, 18, 20, 23, 38, 43, 44, 45, 46, 71, 80, 81, 111, 123, 197, 199, 202
Organismic kingdoms, 24, 27, 28, 32, 34, 38, 44, 54, 101, 110, 197, 199, 203

Organogenesis, 29, 38, 182
Overdeveloped hybrids, 204

P
Parasites, 33, 34, 36f, 37, 40f, 71f, 72, 75f, 78f, 96, 97, 99f, 100, 101, 119, 122, 133, 134, 136, 137, 142, 144, 150, 153, 159, 160, 162, 163, 165, 174, 175, 176–180, 182, 189
Parasitic, 32, 33, 37, 91, 96, 102, 114f, 123f, 132, 134, 135, 140, 142, 191, 203
Persistence, 120, 133, 138, 174–175, 177–178
Persistent, 24, 100, 101, 102, 103, 119–121, 122, 131, 132, 133, 134, 138, 139, 144, 150, 151, 153, 154, 162, 163, 164, 166, 173, 174, 175, 176, 177, 178, 180, 183, 189, 190, 191
Persistent endogenous retroviruses, 150, 183
Persistent exogenous agents, 164
Persistent invaders, 183
Persistent lifestyle, 24, 103, 121, 173, 176, 178
Persistent phages, 120
Phage ecology, 120
Phages, 118, 119, 120, 121, 138, 153, 154, 160, 164, 165, 174
Phenotypic variations, 130, 182
Philosophical Investigations, 16
Philosophy of science, 1, 7, 17, 23, 202
Phylogenetic analyses, 122, 132, 141, 143, 175
Physicalism, 20, 22
Physics, 4, 5, 9, 144, 199, 200, 201, 202
Placental mammals, 129, 140, 142–143
Placental tissue, 142
Planning, 32
Plant
body, 32, 35, 42
neurobiology, 42–44
stem, 31
synapse, 29, 34
Plasmids, 43, 101, 118, 121, 122, 154, 160, 161, 162, 164, 176, 187
Plasmodesmata, 28, 37–39, 98
Position effect variegation, 130
Pragmatic action theory, 19, 23
Pragmatic aspects, 19, 111, 198
Pragmatic rules, 22, 28, 44, 111, 143, 163, 199
Pragmatics, 19, 22, 23, 111, 150, 165, 172, 198
Pragmatic turn, 16–20, 22, 23
Prebiotic assembly of ribonucleotides, 151
Prebiotic RNA copies, 151
Pre-cellular, 129, 135–136, 150, 151, 152, 203
Pre-cellular consortium, 152
Pre-cellular RNA copies, 150–151
Pre-DNA world, 150
Pre-mRNAs, 162, 163, 181
Pre-transcriptional, 171
Priming, 152
Prions, 100
Private language, 17
Processual reality, 6, 7
Prokaryotes, 71, 109, 116f, 118f, 119, 121, 122, 123f, 136, 137, 139, 149, 152, 153, 162, 164, 174, 175, 180, 184
Protein-coding data sets, 153, 164, 165
Protein-coding DNA, 41, 42
Protein meanings, 42, 130, 131, 143, 144, 164, 166, 189
Protoctista, 90, 95, 144

Q
Quorum sensing, 33, 43, 95, 96, 110, 111, 112, 113, 114, 116, 123
Quorum-sensing 'dialects', 116
Quorum sensing pathways, 113

R
Random mixture of nucleotides, 150
Random mutations, 136, 149, 172
Rationalism, 3, 7, 15
Reactive Oxygen Species, 30f, 31, 93
Receptors, 11, 32, 33, 38, 97, 98, 113, 133, 140, 141, 178
Recognition, 33, 37, 43, 68, 76, 96, 113, 117, 121, 122, 133, 141, 142, 157, 159, 172, 178, 181
Recombinase activating genes, 141
Recombination, 24, 40f, 75, 78f, 110, 117, 119, 120, 121, 122, 130, 135, 136, 141, 153, 159, 163, 171, 174, 190
Reef ecosystems, 70, 73, 77, 80, 81
Regulation networks, 31, 122
Regulatory agents, 152, 166
Regulatory elements, 100, 153, 163, 166, 189
Regulatory function, 31, 129, 130, 142, 143, 152, 161, 162, 165, 176, 177, 180, 183
Repair proteins, 120
Repeat elements, 101, 160, 162, 164
Repetitive DNA, 41, 157
Repetitive DNA sequences, 157
Repetitive genomic sequences, 181
Repetitive sequences, 131, 142, 151, 160, 182
Replication, 11, 41, 119, 120, 121, 130, 131, 132, 135, 136, 137, 138, 139, 150, 151, 152, 153, 157, 158, 159–162, 163, 166, 171, 174, 175, 176, 177, 178, 182, 183, 186, 187, 188, 189, 190
Replication system, 121, 132, 135

Response behavior, 103f, 112, 113, 115, 118, 123f
Responses, 28, 29, 30, 31, 32, 33, 34, 39, 41, 68, 72, 76, 95, 97, 99, 100, 102, 142, 197
Restriction, 76, 79, 121, 122, 133, 134, 138, 139, 176, 191
Retroagents, 150, 151, 187
Retrodirection-reading, 150
Retro-editing, 151
Retroelements, 41, 151, 153, 160, 163, 164, 165, 177, 183, 184, 185, 187
Retroplasmids, 118, 121
Retroposons, 100, 102, 130, 140, 150, 153, 159, 160, 161, 164, 174, 176, 177, 182, 184, 185, 186
Reverse transcription, 41, 187
Rhizobia bacteria, 34, 114
Ribonucleoproteins, 152, 153, 159, 180, 183–184
Ribonucleoprotein structures, 152
Ribosome biogenesis, 99
Ribozymatic function, 152
Risc, 176
RNA-based communities, 191
RNA-based life-forms, 153
RNA-chaperones, 153
RNA Code, 150–151
RNA correction, 153
RNA-dependent RNA polymerases, 150, 159, 161
RNA hairpins, 181
RNA helicase A, 179
RNAi, 31, 69, 69f, 92f, 133, 134, 139, 151, 159, 161, 165, 176, 182, 188
RNA interference, 130
RNA population, 163, 188
RNA silencing, 36, 133, 152
RNA stem-loop structures, 152
RNA template, 158, 159, 187
RNA-Transport-Granulat, 179
RNA-viral replicases, 132
RNA-world, 132, 149–154, 174
Root zone, 28, 29, 33, 34, 36, 37, 42
Round-dance, 58
RRNA, 152, 153, 162, 163, 171, 182, 185, 186, 187, 190
Rule-governed sign-mediated interactions, 20, 60, 94f, 123f, 203
Rules, 12, 17, 20, 23, 24, 28, 44, 45, 53, 54, 55, 60, 61, 63, 80, 81, 94f, 111, 116, 117, 123f, 143, 144, 163, 190, 198, 199, 200, 201, 202, 203

S
Satellite RNAs, 152
Scientific communities, 18, 19, 68
Scientific languages, 13, 15–16, 17, 18, 19, 130
Scleractinian corals, 80
Self, 36, 36f, 39, 43, 45f, 68, 75, 75f, 80f, 90, 95, 96f, 97, 99f, 113, 116f, 117, 122, 151, 152, 163, 174, 175, 182, 188
Self-incompatibility, 39
Selfish genetic elements, 172
Selfnonself-identification, 140
Self-replicating RNA species, 24
Semantic aspects, 10, 111, 198
Semantic functions, 28, 45
Semantics, 10, 14, 19, 21, 23, 42, 68, 81, 111, 150, 165, 172, 198, 202
Semiochemicals, 23, 24, 33, 46, 92f, 93, 110, 112f
Semiochemical vocabulary, 30f, 69–70, 91–92, 92f, 109, 112–113, 112f
Semiotic rules, 18, 20, 23, 24, 28, 44, 45f, 80, 80f, 81, 111, 173, 200, 201, 202, 203
Semiotics, 22, 63, 199, 202
Sender-receiver narrative, 9, 15
Sequence, 10, 21, 22, 24, 28, 39, 41, 42, 43, 44, 45, 53, 57, 60, 61, 62, 63, 91, 96, 101, 102, 109, 117, 118, 122, 129, 130, 131, 132, 137, 142, 144, 150, 151, 152, 153, 157, 158–159, 160, 161, 162, 163, 164, 165, 172–175, 178, 179, 181, 182, 184, 187, 188, 189, 190, 191, 198, 199, 200, 202, 203
Sequence editing competences, 152
Serial endosymbiotic theory, 136
Sessile life-style, 132, 133
Sexual isolation, 135
Short interspersed elements, 130
Sickle-dance, 58
Signalling cascades, 91, 98
Signalling pathways, 38, 42, 44, 77, 92, 93, 97, 98, 99, 102, 183, 197
Signal molecules, 28, 29, 30, 31, 39, 45, 68, 69, 72, 78, 81, 91, 98, 100, 110, 112, 113, 114, 123
Signal perception, 102
Signal sequences, 39
Signal transduction processes, 180, 197, 198
Signs, 9, 10, 12, 15, 18, 19, 20, 21, 22, 23, 24, 27, 28, 29, 30f, 32, 44, 45f, 46, 53, 54, 55, 56, 57–58, 60, 61, 62, 63, 64, 68, 69f, 80, 81, 92f, 110, 111, 112f, 114, 123, 151, 172, 173, 198, 199–200, 201, 202

Sign-user, 22, 23, 199
SINEs, 101, 130, 142, 143, 160, 164, 184, 190
Single-stranded RNA viruses, 100, 132, 133, 187
Situational contexts, 12, 28, 42, 69, 111, 115, 123, 144, 173, 188, 198, 199, 200
Small interfering RNA, 31, 102, 180, 181, 182, 189
Small nuclear RNAs, 163, 180, 181, 182, 186
Small nucleolar RNAs, 160, 163, 171, 180, 183, 184–187
Small nucleolus-like organelles, 161
SnoRNAs, 161, 184, 185, 186, 190
SnRNPs, 184, 185, 187
Social coordination, 18
Social experience, 14, 18
Social intelligence, 43
Social interactions, 12, 18, 19, 21, 39, 53, 54, 59
Sociology, 18
Solipsistic, 17, 22, 23, 39
Speech, 9, 12, 14, 16, 19, 23
Speech-act theory, 19
Spindle microtubules, 157
Splice junctions, 181
Spliceosomes, 163, 171, 181, 182, 191
Sporulation, 110, 112, 113, 116, 123
SsRNA-parasites, 150
Stress, 7, 28, 30, 38, 41, 42, 68, 72, 73, 74, 76, 91, 93, 95, 98, 102, 112, 120, 144, 191
Structural code, 21, 111, 198
Subject of knowledge, 19
Subviral-lineages, 152
Superficial grammar, 164
Suppression of transposition, 130
Swarm, 53, 54, 55, 56, 57, 58, 61, 62, 63, 112, 115, 163, 174, 191
Symbiodinium, 74, 79, 120, 176
Symbiogenesis, 34, 35, 115
Symbiology, 82, 110, 117, 201
Symbionts, 28, 39, 67, 68, 70, 77, 79, 81, 97
Symbiosis, 35, 77, 93, 95, 100, 115, 176
Symbiotic communication, 32, 71
Symbiotic interactions, 24, 34, 35–36, 73, 74, 91, 115, 174
Symbiotic relationships, 34, 119
Symbols, 9, 15, 20, 54, 63, 172, 199, 200
Symplastic signalling, 38
Syncytin, 143
Syntactic rules, 28, 44, 81, 111, 188, 199
Syntax, 9, 10, 13, 14, 15, 16, 19, 21, 23, 28, 46, 81, 111, 120, 123, 143, 144, 150, 153, 154, 158–159, 163, 165, 166, 172, 179, 181, 188, 189, 198, 202

T

Tandem repeats, 41
TART, 159, 160
Telomerases, 24, 157–166, 185, 186, 187
Telomere repeats, 152, 157, 158, 161, 162, 166, 174, 175
Telomeres, 24, 138, 150, 152, 153, 157–166, 174, 175, 185, 186, 187, 189
Textbook conviction, 42, 149
Text-formatting, 173
Theory of biocommunication, 20, 23
Theory-language, 13
Theory of levels, 6
Tissues, 20, 23, 28, 29, 33, 37, 38, 72, 73, 75, 76, 77, 78, 79, 91, 94, 119, 132, 140, 142, 143, 176, 178, 180, 199, 202
Transcription, 31, 38, 41, 79, 98, 99, 102, 110, 113, 118f, 119, 121, 122, 130, 131, 136, 138, 139, 142, 143, 152, 153, 160, 161, 162, 166, 171, 172, 174, 175, 177, 178, 180, 181, 182, 183, 184, 185, 186, 187, 189, 190, 191, 198
Transduction, 43, 81, 98, 121, 131, 180, 197, 198
Transfer RNA consortium, 187–188
Transformation, 117, 121, 166
Translation, 6, 13, 31, 41, 99, 119, 122, 131, 136, 138, 139, 152, 162, 164, 172, 174, 175, 176, 178, 180, 181, 183, 185, 187, 189, 190, 191, 198
Trans-organismic communication, 24, 63, 95
Transposable elements, 41, 163, 182
Transposon, 41, 101, 160, 181
Transposon control, 181
Trans-specific signalling, 117
Tremble-dance, 58
TRNA, 152, 153, 162, 163, 171, 181, 182, 185, 187–188, 190, 191
Trophoblast, 142, 143
Ty1, 101, 160

U

Umwelt, 22, 203
Understanding, 2, 8, 9, 10, 12, 15, 17, 19, 20, 23, 24, 27, 62, 63, 67, 68, 74, 98, 144, 165, 197, 198, 202, 203, 204
Unencapsulated RNA molecules, 150
Universal grammar, 111, 198
Universal syntax, 9, 10, 13, 16, 21

V
Vertical colonisation, 109, 119
Vibration-dance, 58
Viral agents, 129, 139, 142, 164, 175–180, 183
Viral colonization, 122, 134, 139, 142, 144, 165
Viral competences, 118, 119, 121, 122, 152, 164, 165, 166, 174, 178, 179
Viral infection, 101, 102, 118f, 119, 121, 123f, 131, 133, 154, 166, 181
Viral inventions, 122, 139
Viral-like agents, 24, 176
Viral persistence, 133
Viral soup of archaic oceans, 151
Viroids, 36, 39, 78, 99f, 103f, 152
Viruses, 24, 35, 36, 39, 40, 40f, 45f, 71, 72, 73, 78, 78f, 79, 80f, 95, 99f, 100–102, 103, 103f, 109, 118f, 119, 120, 121, 122, 129–145, 149, 150, 151, 152, 153, 159, 161, 163, 164, 165, 173, 174, 175, 176, 177, 178, 179, 181, 182, 183, 185, 187, 189, 190, 203
Virus-first, 150, 174, 191
Volatiles, 32, 33, 36, 37, 42, 57, 96

W
Waggle dances, 56, 57, 58, 62, 63
Watson-Crick basepairing, 150
Wavelength, 32

Y
Yeast, 95, 100, 101, 130, 158, 184, 187

Lightning Source UK Ltd.
Milton Keynes UK
05 March 2010

151001UK00007B/41/P

9 789048 133185